Dynamics of Plate Interiors

Geodynamics Series

The Final Reports of the International Geodynamics Program sponsored by the Inter-Union Commission on Geodynamics.

Dynamics of Plate Interiors

Edited by A. W. Bally
P. L. Bender
T. R. McGetchin
R. I. Walcott

Geodynamics Series
Volume 1

American Geophysical Union
Washington, D. C.
Geological Society of America
Boulder, Colorado
1980

Final Report of Working Group 7, Geodynamics of
Plate Interiors, coordinated by R. D. Russell on
behalf of the Bureau of Inter-Union Commission
on Geodynamics

ISBN # 87590-508-0

American Geophysical Union, 2000 Florida Avenue, N.W.
  Washington, D. C. 20009

Geological Society of America, 3300 Penrose Place; P. O. Box 9140
  Boulder, Colorado 80301

Printed in the United States of America

CONTENTS

After a decade of intense and productive scientific cooperation between geologists, geophysicists and geochemists the International Geodynamics Program formally ended on July 31, 1980. The scientific accomplishments of the program are represented in more than seventy scientific reports and in this series of Final Report volumes.

The concept of the Geodynamics Program is a natural successor to the Upper Mantle Project developed during 1970 and 1971. The International Union of Geological Sciences (IUGS) and the International Union of Geodesy and Geophysics (IUGG) then sought support for the new program from the International Council of Scientific Unions (ICSU). As a result the Inter-Union Commission on Geodynamics was established by ICSU to manage the International Geodynamics Program.

The governing body of the Inter-Union Commission on Geodynamics was a Bureau of seven members, three appointed by IUGG, three by IUGS and one jointly by the two Unions. The President was appointed by ICSU and a Secretary-General by the Bureau from among its members. The scientific work of the Program was coordinated by the Commission, composed of the Chairmen of the Working Groups and the representatives of the national committees for the International Geodynamics Program. Both the Bureau and the Commission met annually, often in association with the Assembly of one of the Unions, or one of the constituent Associations of the Unions.

Initially the Secretariat of the Commission was in Paris with support from France through BRGM, and later in Vancouver with support from Canada through DEMR and NRC.

The scientific work of the Program was coordinated by ten Working Groups.

WG 1 Geodynamics of the Western Pacific-Indonesian Region
WG 2 Geodynamics of the Eastern Pacific Region, Caribbean and Scotia Arcs
WG 3 Geodynamics of the Alpine-Himalayan Region, West
WG 4 Geodynamics of Continental and Oceanic Rifts
WG 5 Properties and Processes in the Earth's Interior
WG 6 Geodynamics of the Alpine-Himalayan Region, East
WG 7 Geodynamics of Plate Interiors
WG 8 Connections Between Oceanic and Continental Structures
WG 9 History and Interaction of Tectonic, Metamorphic and Magmatic Processes
WG 10 Global Syntheses and Paleoreconstruction

These Working Groups held discussion meetings and sponsored symposia. The papers given at the symposia were published in a series of Scientific Reports. The scientific studies were all organized and financed at the national level by national committees even when multinational programs were involved. It is to the national committees, and to those who participated in the studies organized by those committees, that the success of the Program must be attributed.

Financial support for the symposia and the meetings of the Commission was provided by subventions from IUGG, IUGS and UNESCO.

Information on the activities of the Commission and its Working Groups is available in a series of 17 publications: Geodynamics Reports, 1-8, edited by F. Delany, published by BRGM; Geodynamics Highlights, 1-4, edited by F. Delany, published by BRGM; and Geodynamics International, 13-17, edited by R. D. Russell. Geodynamics International was published by World Data Center A for Solid Earth Geophysics, Boulder, Colorado 80308, USA. Copies of these publications, which contain lists of the Scientific Reports, may be obtained from WDC A. In some cases only microfiche copies are now available.

The work of the Commission will be summarized in a series of Final Report volumes. The Final Report volumes were organized by the Working Groups. This volume from Working Group 7 on The Dynamics of Plate Interiors is the first to be published. The Final Report volumes represent in part a statement of what has been accomplished during the Program and in part an analysis of problems still to be solved. At the end of the Geodynamics Program it is clear that the kinematics of the major plate movements during the past 200 million years is well understood, but there is much less understanding of the dynamics of the processes which cause these movements.

Perhaps the best measure of the success of the Program is the enthusiasm with which the Unions and national committees have joined in the establishment of a successor program to be known as: Dynamics and evolution of the lithosphere: The framework for earth resources and the reduction of the hazards.

To all of those who have contributed their time so generously to the Geodynamics Program we tender our thanks.

C. L. Drake, President ICG, 1971-1975

A. L. Hales, President ICG, 1975-1980

Members of Working Group 7:

D. E. Ajakaiye
E. V. Artyushkov
A. W. Bally

J. J. Bigarella
J. D. Boulanger
P. J. Burek
B. J. Collette
A. Dudek
H. Faure
D. I. Gough
S. L. N. Kailasam
A. P. Kapitsa
W. W. Kaula
E. A. Lubimova
V. A. Magnitsky
T. R. McGetchin
V. D. Nalivkin
D. A. Pretorius
D. L. Turcotte
J. J. Veevers
P. Vyskocil
R. I. Walcott
J. T. Wilson
V. Zorin

The focus of the International Geodynamics Project, 1970-1979, was the movements of the surface and upper part of the earth's interior and it was recognized that most of the deformation occurs along narrow belts between the lithospheric plates. Also important to understanding earth processes were those motions, primarily vertical that occurred within the plates, remote from plate boundaries. For this reason one of the 10 working groups set up in 1971 was working group 7 with the title "Eperogenic movements of regional extent" under the chairmanship of Dr. J. Tuzo Wilson. In 1974, after Dr. Wilson resigned following his retirement as Principal of Erindale College, University of Toronto, the Bureau of the Inter-Union Commission on Geodynamics appointed Dr. R. I. Walcott as chairman and the name of his working group was changed to "Dynamics of Plate Interiors". The objective of its programme was to determine the nature and origin of the dynamics of the more stable regions of the earth.

But the Geodynamics Project did more than merely identify problems. The very reason for its existence was to bring together and encourage an interdisciplinary attack to those problems - to join those skilled with inversion and other modelling techniques with those whose observations provided the information necessary for adequate modelling, to associate those who were developing instruments for the precise measurement of relative motion by modern geodetic techniques with those with knowledge of past movements in geological time in an endeavour to provide a single line of interdisciplinary investigation.

This requirement of an interdisciplinary approach both sharpened the focus and limited the nature of the problems to be investigated. At the meeting of the Inter-Union Commission on Geodynamics in Grenoble 1975, four areas of study were identified by the working group that promised to respond to an interdisciplinary approach. These were Quaternary Vertical Movements, Vertical Movements from the Stratigraphic Record, History and Mechanism of Plateaux Uplift and the Instrumental Measurement of the Deformation of Plate Interiors.

The principal mechanism by which the working group carried out its task was to encourage the organisation of symposia on these facets of the problem. Accordingly the following symposia were held.

1. Regional Stresses and Models for Epeirogeny, Zurich 1974, convened by R.I. Walcott within the Recent Crustal Movements Symposia.

2. Sedimentary Basins of the Continental Margin and Craton, Durham 1976, organized by M.H.P. Bott as a joint Working Group 7 and 8 symposia.
3. Earth Rheology and Quaternary Isostatic Movements, Stockholm 1977, organised by N.A. Mörner.
4. Mode and Mechanism of Plateau Uplifts, Flagstaff 1978, organised by T.R. McGetchin.
5. Dynamics of Plate Interiors, Canberra 1979, convened by R.R. Walcott as part of the ICG symposia during the IUGG General Assembly.

In each of the symposia geophysicists, geodesists, geologists, and geomorphologists - earth scientists of a wide variety of persuasions and nationalities - took part and were made aware of the different opinions and information of those working in the same field but with different backgrounds and skills.

Papers from Symposium 1 appeared in the special issue of Tectonophysics (Volume 29, December 1975) on Recent Crustal Movements (editors, N. Pavoni and R. Green); from Symposium 2, in another special issue of Tectonophysics (Volume 36) titled "Sedimentary Basins of the Continental Margins and Cratons" (editor, M.H.P. Bott); and from Symposium 3, in the book "Earth Rheology, Isostasy and Eustasy" (editor, Nils-Axel Morner), John Wiley and Sons, 599 pp. 1980. A report of Symposium 4 appeared in EOS, the publication of the American Geophysical Union, and this final report of the working group. Most of the papers of Symposium 5 have either already been published or will shortly be so in the usual journals.

This, the final report of the working group, brings together a number of papers on the general topic of the "Dynamics of Plate Interiors". They are grouped into four parts, each reflecting one of the facets of the activity of the working group identified in 1975. Some of the papers are written by members of the working group but most were solicited from authors who bring a special expertise to the subject. Most of the subject matter of these papers already appeared in published form, in the publications referred to above, or in specialized journals. However, what is new, and what is intended by the working group in presenting this report, is a change in the audience to whom the papers are directed.

One of the major difficulties in working across disciplinary boundaries is that the specialized nature of the language and style of presentation, while facilitating ease and accuracy of presentation, makes the extraction of information difficult for those of other disciplines. For example, modelling of basin subsidence or postglacial rebound is usually given by mathematical expressions in a language commonly not well understood by geologists involved in basin studies or by geomorphologists who observe uplifted beaches. Or again, stratigraphic data are presented by use of a wide variety of formation names or stratigraphic units with a precise and restricted meaning that presents the physicist with some difficulty in extracting useful information on, say, the relative vertical motion of the basement from stratigraphic data on the cover rocks. The authors of these papers were therefore requested to identify as their audience those in quite different disciplines to their own who would be interested in their information. We feel that in general the authors have been very successful in meeting this request.

The parts are quite unequal in size, partly reflecting the degree of enthusiasm brought to the project by those involved, but primarily due to the relative importance of these facets to the overall problem.

The largest, Part I, consists of 9 papers on Vertical Movements from the Stratigraphic Record. This was identified very early in the proceedings of the working group as by far the most important aspect. Here, there is a truly vast amount of information on the past history of the stable regions of the earth, particularly of the continental regions. With seismic data, drilling and stratigraphic analysis, it is possible, at least in principle, to determine the relative vertical motion of large parts of the earth's surface, and by satisfactorily modelling in terms of conceptually simple processes valuable insights of the behaviour of the earth could be gained.

Part II is a summary of the proceedings of the symposium held at Flagstaff and is written by four of the participants. Dr. T.R. McGetchin, the senior author and the organiser of the symposium has since died and the earth scientists have lost one of their most able and distinguished colleagues who contributed importantly to the work of the Geodynamics project.

The paper by W.R. Peltier in Part III admirably fulfills the request to generalise his work in the interpretation of postglacial rebound data.

The four papers of Part IV cover some of the more important aspects of instrumental measurements on plate interiors including levelling, stress estimates and recent advances in geodetic techniques. These are very rapidly developing fields and in the next decade we expect to see an even greater contribution of useful data to the understanding of processes within the continental parts of the plates.

BASINS AND SUBSIDENCE - A SUMMARY

A. W. Bally

Shell Oil Company, P. O. Box 481, Houston, Texas 77001

Introduction

A wide variety of perspectives - properly representing the somewhat disjointed state of the art - is contained in this collection of papers on basin evolution and subsidence. These contributions can be easily grouped in papers concerned with geophysical models of basin subsidence and papers of a dominantly descriptive nature that - with some exceptions - do not make a deliberate effort to quantitatively test the subsidence and uplift history with one or more geophysical models. Although all authors made a diligent effort to provide their contributions on time, the final publication of this volume was unfortunately delayed. Therefore, this summary will also try to call attention to some relevant recent papers that were published after the authors submitted and edited their manuscripts. For another review of the genesis of platform basins, see Sleep et al., 1980.

Geophysical Models

In this volume, Turcotte offers a general introduction to evolutionary models for sedimentary basins; Bott illustrates the applicability of such models to passive margins; finally, Artyushkov et al., discuss the possible effects of differentiates rising from the lower mantle to spread in traps at the base of moving lithospheric plates. In some cases, such spreading even extends under the crust. In principle, the following main subsidence mechanisms have been proposed:

1. Subsidence due to sediment loading. According to Bott and also other authors, subsidence due solely to this mechanism cannot reasonably exceed two to three times the initial water depth and thus may not apply to very thick sequences of shallow water deposits; but, as stated by Bott and Turcotte, sediment loading may well have a significant influence on the subsidence of thick sequences deposited in a deep-water or a continental rise environment. Consequently, other processes or "driving forces" have to be invoked to explain what

C. Keen (1979) calls "tectonic subsidence".

2. Isostatic subsidence due to cooling of a previously heated lithosphere and the associated increase in density. Heating of the lithosphere and its crust results in uplift and crustal thinning due to surface erosion. Later, cooling would cause subsidence of such an attenuated crust. However, the amounts of uplift and surface erosion required to explain the origin of deep sedimentary basins (i.e., in excess of 5 km) are unreasonably large; and, therefore, additional mechanisms are needed (see Royden et al., 1980).

3. Subsidence due to cooling may also be the consequence of simple lithospheric stretching (McKenzie, 1978; Royden et al., 1980; Sclater and Christie, 1980). Such stretching may be accomplished by listric normal faulting in the upper crust combined with ductile necking of the lower crust and/or upper mantle.

4. Not discussed in this series of papers is a mechanism proposed by Royden et al. (1980), whereby sedimentary basins and passive continental margins may be formed by cracking of continental lithosphere and the intrusion of ultrabasic dikes and/or diapirs. This model is related to the ones described later under 5 and 6, but it envisages the process to be short-lived and limited to the formation of graben systems, that may initiate sedimentary basins.

5. A density increase of lower crustal or lithospheric rocks due to gabbro-eclogite phase changes or else metamorphism may also lead to isostatic subsidence. It has, however, been difficult to demonstrate that such a process has occurred on a large scale in nature (see, also, Sleep et al., 1980). The effect of transformation of metastable phases to denser stable phases at lower crustal conditions on passive margins has been discussed by Neugebauer and Spohn (1978).

6. The two preceding mechanisms may be viewed as special cases of partial "oceanization". However, complete basification or "oceanization" of the crust is believed to be caused by a supply of ultrabasic material from the mantle or deeper. In this process, the top layer of the crust becomes basic, while the lower crust

TABLE 1 - BASIN CLASSIFICATION

1. **BASINS LOCATED ON THE RIGID LITHOSPHERE, NOT ASSOCIATED WITH FORMATION OF MEGASUTURES**

    11. Related to formation of oceanic crust

        111. Rifts

        112. Oceanic transform fault associated basins

        113. Oceanic abyssal plains

        114. Atlantic-type passive margins (shelf, slope & rise) which straddle continental and oceanic crust

            1141. Overlying earlier rift systems
            1142. Overlying earlier transform systems
            1143. Overlying earlier Backarc basins of (321) and (322) type

    12. Located on pre-Mesozoic continental lithosphere

        121. Cratonic basins

            1211. Located on earlier rifted grabens
            1212. Located on former backarc basins of (321) type

2. **PERISUTURAL BASINS ON RIGID LITHOSPHERE ASSOCIATED WITH FORMATION OF COMPRESSIONAL MEGASUTURE**

    21. Deep sea trench or moat on oceanic crust adjacent to B-subduction margin

    22. Foredeep and underlying platform sediments, or moat on continental crust adjacent to A-subduction margin

        221. Ramp with buried grabens, but with little or no blockfaulting
        222. Dominated by block faulting

    23. Chinese-type basins associated with distal blockfaulting related to compressional or megasuture and without associated A-subduction margin

3. **EPISUTURAL BASINS LOCATED AND MOSTLY CONTAINED IN COMPRESSIONAL MEGASUTURE**

    31. Associated with B-subduction zone

        311. Forearc basins
        312. Circum Pacific backarc basins

            3121. Backarc basins floored by oceanic crust and associated with B-subduction (marginal sea sensu stricto).
            3122. Backarc basins floored by continental or intermediate crust, associated with B-subduction

    32. Backarc basins, associated with continental collision and on concave side of A-subduction arc
        321. On continental crust or Pannonian-type basins
        322. On transitional and oceanic crust or W. Mediterranean-type basins

    33. Basins related to episutural megashear sytems

        331. Great basin-type basin
        332. California-type basins

Table 1. Basin Classification (after Bally and Snelson, 1980, with permission of Can. Soc. Petrol. Geologists).

becomes ultrabasic and acquires a density equal to or even greater than the mantle (Beloussov, 1968). According to Artyushkov, such a process is implausible, because a mixture of bodies of usual crustal and mantle composition will always remain lighter than mantle matter and, consequently, cannot sink into it.

7. Bott, in this volume, reviews his explanation for passive margin subsidence and crustal thinning due to creep of a ductile middle and lower crust toward the ocean. Such creep is caused by unequal topographic loading across passive continental margins.

Vetter and Meissner (1979) modified Bott's concepts and divided the lithosphere into an upper brittle layer and a lower layer in which high temperature creep occurs. On passive margins high density oceanic lithosphere will creep under the adjacent continent, a process which eventually may initiate subduction and the conversion from passive margin to an active margin. In their view - and in contrast to Bott - these authors feel that the viscosity of the middle crust is too high to permit creep.

8. Sloss and Speed (1974) and Sloss (1980) visualize variations of the rate of flow of melt from continental to oceanic asthenosphere. In their model, trapping of melt beneath continents causes uplift, while the draining of melt to oceanic asthenospheric circulation would cause continental deflation and subsidence that would also coincide with accelerated sea-floor spreading.

9. Artyushkov, in this volume, relates uplift and subsidence events to periodically upwelling differentiates that are formed at the core-mantle boundary and that spread below the lithosphere.

For a review of the lithospheric rheologic characteristics that are relevant to basin subsidence models, the reader is referred to Beaumont (1978), Beaumont and Sweeney (1978) and Beaumont (1979). Particularly in the last publication, the author states that it is difficult to decide among competing rheologies (Elastic, Elastic-Perfectly plastic, Visco-elastic and Power Law Creep) or various combinations of these properties. Therefore, more information on initiating mechanisms and on present and past lateral changes in lithospheric properties will be needed to obtain a more complete understanding of lithosphere rheology.

Basin Classification

An explanation of the great variety of basin types may require different combinations of the above-mentioned mechanisms and possibly yet other mechanisms that so far have not been proposed. From a geologic perspective, it is on the other hand desirable to classify basin types if only to get an overview of the large number of individually characteristic basins that occur in the world. This author naturally prefers his own classification (Bally, 1976; Bally and Snelson, 1980), but quite obviously such classifications have a degree of arbitrariness and can easily be replaced by alternate systems. The classification (Table 1) is meant to replace the old geosynclinal classification, because the latter does not adequately reflect advances in plate tectonics. Fig. 1 provides a megatectonic framework, which is used as background for a classification. Figs. 2, 3, 4 show the distribution of three major basin families, i.e., basins located on the rigid lithosphere, perisutural basins and episutural basins. Fig. 5 shows that for North America, the different basins occur side by side on a single continent. From a geological

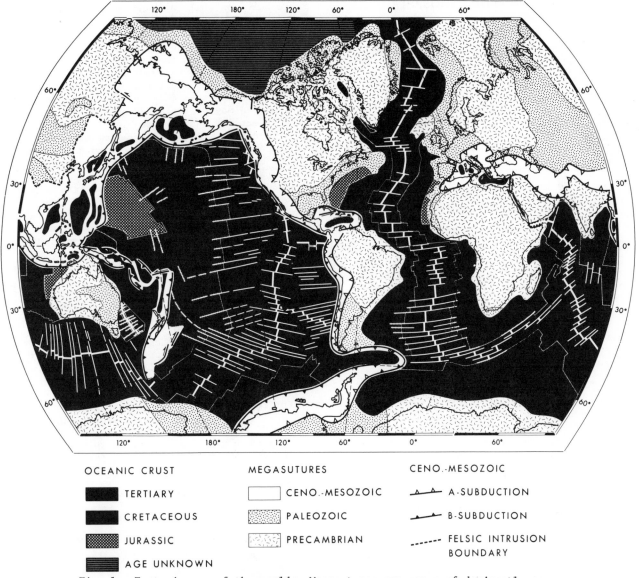

Fig. 1. Tectonic map of the world: Megasutures are zones of dominantly non-rigid orogenic deformation and sedimentary basins included within such zones. A-subduction (after Ampferer) involves the limited subduction of continental lithosphere; the symbol corresponds to the boundary of a folded belt facing and partially overthrusting continental crust. B-subduction (after Benioff) is subduction of oceanic lithosphere with folded belts facing oceanic crust. The felsic intrusion boundary separates areas of Mesozoic and Cenozoic igneous activity in China from areas that are not so affected (after Bally and Snelson, 1980, with permission of Can. Soc. Petrol. Geologists).

perspective, a different genesis may be expected for the different basin types and some of their subdivisions.

In the following, the classification will be used as a guide to review the applicability or non-applicability of the many proposed subsidence mechanisms to some of the more important basin classes.

Concerning the Applicability of Various Geophysical Models to Different Basin Types

The following observations are based on the papers contained in the Basin and Subsidence section. For the sake of continuity, we have refrained from citing the specific observations and thoughts of the authors of this volume and

BASINS ON RIGID LITHOSPHERE

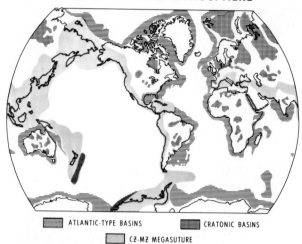

Fig. 2. Basins on rigid lithosphere. These are further subdivided on Table 1 (after Bally and Snelson, 1980, with permission of Can. Soc. Petrol. Geologists).

of many other publications; we do, however, cite some recent publications that were not available when our authors submitted their papers.

Atlantic-type (passive) continental margins typically show a lower faulted section corresponding to an early rifting event and an upper sequence of sediments that correspond to a drifting phase. The two sections are often separated by one or more widespread unconformities. The oldest sediments of the upper sequence are typically of about the same age as the oldest adjacent oceanic crust.

Atlantic-type margins are well explained by a combination of: (1) Thermal origin, due either to uplift over upwelling deeper asthenospheric material, or else due to simple stretching of the continental lithosphere. The upper brittle layer of the stretched lithosphere may extend along listric normal faults that sole out in the lower crust or upper mantle. (2) Injection of basic intrusive material. (3) Cooling associated with advanced ocean spreading and modified by sediment loading; and (4) creep in the middle and lower crust and possibly ocean toward continent creep in the underlying asthenosphere.

The ultimate causes of the thermal origin are not easily determined. The effect of lithospheric cooling and the amplified subsidence due to sediment loading can be computed, but it remains difficult to differentiate effects due to sediment loading from the more hypothetical effects of creep in the lower crust, or of asthenospheric creep.

Recent studies by Steckler and Watts (1978), Keen (1979) and Watts and Steckler (1979) on the subsidence of the North Atlantic margin were based on well data and reflection seismic studies. The subsidence reconstructions of these authors included proper corrections for compaction (i.e., the layers were "decompacted") and for water depth changes as indicated by paleoenvironmental bathymetric analysis. According to Keen, in the order of 55 to 65 percent of the subsidence on the Canadian Atlantic margin may be explained by sediment loading; the effects of eustatic sea level changes on subsidence is small; and the remainder of the subsidence which the above-mentioned authors refer to as "tectonic subsidence" is by all authors found to be explainable as isostatic response to thermal cooling of the lithosphere. Watts and Steckler then attempt to isolate the "eustatic" effects by assuming that tectonic subsidence is indeed thermal and by least square fitting of an exponential curve to the subsidence data. They conclude that the maximum rise of sea level during late Cretaceous is not likely to exceed 150 m.

Studies by Montadert et al (1979) across the Atlantic-type margin of the northern Gulf of Biscay were particularly useful in differentiating the effects of a submarine rifting event from later subsidence due to cooling. Active rifting in the northern Gulf of Biscay is accompanied by 10 to 15 percent stretching along listric normal faults of the upper brittle crust. This amount of stretching is not sufficient to correspond to a thinning of about three km of the lower ductile crust. Therefore, Montadert et al., postulate that the ductile part of the lower crust is thinned by creep in response to tension in the continental

PERI-SUTURAL BASINS

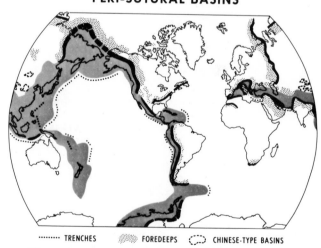

Fig. 3. Perisutural basins, adjacent to subduction boundaries but located on rigid lithosphere. These are further subdivided on Table 1 (after Bally and Snelson, 1980, with permission of Can. Soc. Petrol. Geologists).

plate. Post-rifting (Aptian and later) subsidence decreases exponentially and thus may be explained by an isostatic readjustment of the crust due to cooling. The actual amount of subsidence decreases continuously from the ocean-continent boundary (at about 4000 m) toward the shelf break. Thus, for any point on the margin, the subsidence versus time curve is an exponential, the time constant of which increases with depth.

Recently LePichon and Sibuet (1980) re-examined the Gulf of Biscay area in the light of McKenzie's (1978) stretching model. They conclude that "the amount of brittle stretching in the upper 8 km of continental crust reaches a maximum value of about 3 and is equal to the stretching required to thin the continental crust and presumably the lithosphere".

Cratonic basins appear to vary widely among themselves. For a large number of cratonic basins we have only a very dim picture of their early history. But it is fair to state that where we have the data, most cratonic basins appear to be underlain by more or less complex rift systems. Consequently, one is tempted to apply a passive margin-type thermal model for the genesis of cratonic basins, that is, a model where a thermal event initiates the rifting and subsequent cooling would lead to an exponential decrease in the subsidence rate.

EPI-SUTURAL BASINS

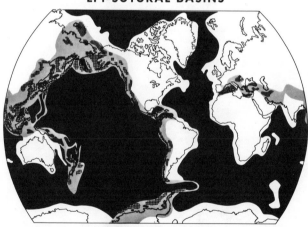

Fig. 4. Episutural basins located on Cenozoic-Mesozoic megasuture. These are further subdivided on Table 1. A small number of preserved Paleozoic episutural basins (e.g., Gulf of St. Lawrence, Sidney Basin) are not shown (after Bally and Snelson, 1980, with permission of Can. Soc. Petrol. Geologists).

Unfortunately, in the case of the Michigan Basin, the timing of the presumed rifting event is much too early to justify at least two heating events that would be required to explain the observed subsidence. The introduction of multiple heating events for which there is little physical evidence is not particularly desirable (Sleep et al., 1980).

For the North Sea, Sclater and Christie (1980) have made a careful study using seismic reflection lines and well data. They have computed subsidence after proper corrections for the effects of compaction and the depth of deposition of the sediments. These authors demonstrate that following earlier crustal extension during the Permian and Triassic, respectively, the main extensional phase occurred in Mid-Jurassic through Late Cretaceous, mostly in (say, between 50-75 kms) the Viking graben. Thermal relaxation of the asthenosphere following extension caused general subsidence and the formation of a saucer shaped basin during Upper Cretaceous and Tertiary times. In other words, the origin of the North Sea basin may be explained in terms of cooling that follows a major stretching event.

For most other cratonic basins, the information regarding initiation and cessation of rifting and subsidence rates is not precise enough to test any of the proposed geophysical models or combinations of models.

Perisutural basins have been subdivided on Table 1, and the most important ones are shown on Fig. 3.

The subsidence of *deep sea trenches* is now generally related to subduction and explained in terms of a cooling oceanic lithosphere, flexural bending and loading by an island arc and its accretionary wedge on one side of that lithosphere (Watts and Talwani, 1974; Parsons and Molnar, 1976; Caldwell et al., 1976, 1979).

Surprisingly, quantitative studies of the subsidence of *foredeeps* and corresponding models have become available only very recently. The development of some foredeeps has been reviewed by Briggs and Roeder (1975), Roeder (1980) and Bally and Snelson (1980). Sections across typical foredeeps are shown on Fig. 7. The examples selected indicate the great importance of these basins for hydrocarbon exploration. From a geological perspective, it would appear that many foredeeps are underlain by a platform sequence of a passive margin type, i.e., with an early rifting history and narrow graben system that may be oriented at relatively high angles to the trends of the adjacent fold belts. Such graben systems are overlain by post-rifting platform sequences. These in turn are overlain by wedges of clastic sediments that thicken towards the adjacent mountain ranges and contain much of the detritus derived from the mountain ranges. In a first approximation, Roeder et al. (1978) illustrated the subsidence history of the Appalachian foredeep (Fig. 8).

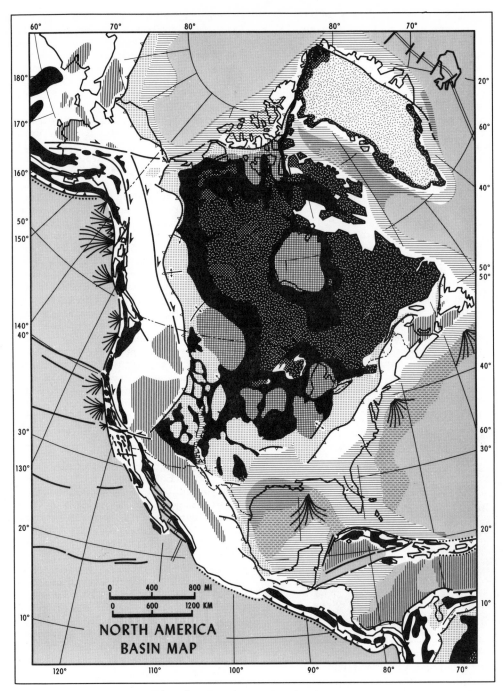

Fig. 5.   North America basin map.

The clastic wedges of foredeeps are frequently separated from the platform sequence and underlain by major supraregional unconformities. These show mappable subcrop patterns that suggest supraregional warping preceding the deposition of the clastic wedge.

The pre-Kaskaskia subcrop shown on Fig. 9a may reflect continent-wide warping immediately preceding the deposition of the clastic wedge related to Devonian tectogenesis in the Appalachian and Innuitian folded belts of North America. The complex pre-Cretaceous subcrop (Fig. 9b) reflects structural events immediately preceding the deposition of the clastic wedge that formed as a consequence of Cretaceous tectogenic events in the Western Cordillera.

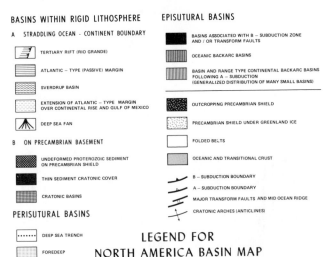

BASINS WITHIN RIGID LITHOSPHERE

A STRADDLING OCEAN - CONTINENT BOUNDARY

TERTIARY RIFT (RIO GRANDE)

ATLANTIC – TYPE (PASSIVE) MARGIN

SVERDRUP BASIN

EXTENSION OF ATLANTIC – TYPE MARGIN
OVER CONTINENTAL RISE AND GULF OF MEXICO

DEEP SEA FAN

B ON PRECAMBRIAN BASEMENT

UNDEFORMED PROTEROZOIC SEDIMENT
ON PRECAMBRIAN SHIELD

THIN SEDIMENT CRATONIC COVER

CRATONIC BASINS

PERISUTURAL BASINS

DEEP SEA TRENCH

FOREDEEP

EPISUTURAL BASINS

BASINS ASSOCIATED WITH B – SUBDUCTION ZONE
AND / OR TRANSFORM FAULTS

OCEANIC BACKARC BASINS

BASIN AND RANGE TYPE CONTINENTAL BACKARC BASINS
FOLLOWING A – SUBDUCTION
(GENERALIZED DISTRIBUTION OF MANY SMALL BASINS)

OUTCROPPING PRECAMBRIAN SHIELD

PRECAMBRIAN SHIELD UNDER GREENLAND ICE

FOLDED BELTS

OCEANIC AND TRANSITIONAL CRUST

B – SUBDUCTION BOUNDARY

A – SUBDUCTION BOUNDARY

MAJOR TRANSFORM FAULTS AND MID OCEAN RIDGE

CRATONIC ARCHES (ANTICLINES)

LEGEND FOR
NORTH AMERICA BASIN MAP

Fig. 6. Legend for the North America map. The Atlantic-type or passive margin basins of North America are related to the opening of the Atlantic and Arctic Oceans during Mesozoic-Cenozoic times; the cratonic basins are dominantly filled with Paleozoic sediments and may or may not display an early Paleozoic rifting event; the foredeeps of the eastern and southern U.S. and the Canadian Arctic are associated with the adjacent Paleozoic fold belts; the foredeeps of western North America are associated with the formation of the adjacent Western Cordillera; the dominantly marine episutural basins of the West Coast and Alaska are either associated with backarc spreading or else with the formation of the San Andreas fault system; the dominantly continental basins of the Basin and Range province are linked with the formation of a major Cenozoic shear system that was superposed on the Western Cordillera of the U.S. and Mexico. Only the generalized distribution of a large number of small basins is shown; the backarc basins of the Caribbean are formed either by backarc spreading (Yucatan Basin) or by capturing of oceanic fragments (Colombian and Venezuelan Basins).

To explain convincingly and to model the subsidence of the deeper rift and platform portion of a foredeep in terms of an Atlantic-type model, more precise timing of the rifting interval and a reconstruction of the subsidence rates are needed. The subsidence of the foredeep clastic wedge may well be explained by a combination of flexural bending and one-sided loading by thrust sheets on a basement ramp that dips under the adjacent mountain range; any additional tectonic subsidence effects would need to be explained in terms of deep crustal flow, loading that occurs deeper in the lithosphere or possibly the crustal delamination under the adjacent uplifted mountains (Bird, 1978).

Beaumont (1980) quantitatively models the subsidence of the clastic wedge of a foredeep in terms of regional isostatic adjustment of the lithosphere under the mass load of the adjacent migrating fold-thrust mountain belt. Thus, the foredeep is due to downward flexure of the lithosphere by the thrust belt; an additional depression is caused by the infilling of mountain derived sediment. Using the Alberta foredeep of Western Canada as a test, Beaumont shows that a viscoelastic rheology for the lithosphere provides the best agreement between model and observation.

In the classification of Table 1, there are two variations of episutural basins, the *block-faulted foredeep basins* (e.g., Wyoming Tertiary basins) and the *Chinese-type basins*, for which no quantitative models have been produced. Both basin types have faulted margins, and the geometry and nature of these marginal faults is the subject of much debate. In the Wyoming and Colorado Rockies, one school claims that the bounding faults steepen with depth (for a review, see Matthews III, 1978); others, however, are swayed by convincing geophysical evidence (Smithson et al., 1978, 1979; Brewer et al., 1980) that at least some of the faults are reverse faults that "sole" or flatten in the lower crust or upper mantle. The latter interpretation would suggest that the Tertiary subsidence in the adjacent basin is in essence flexure due to loading of the lithosphere by a crystalline basement thrust sheet.

A similar mechanism may be invoked for the subsidence of Chinese-type basins (for a map showing the distribution of these basins, see Bally, 1980). In China, as in the Colorado-Wyoming-Utah Rocky Mountains, more geophysical data and careful subsidence studies are needed.

Episutural basins are subdivided into subgroups on Table 1, and the types associated with B-subduction zones (Bally and Snelson, 1980) are shown on Fig. 10.

In general, the subsidence of *forearc basins* (basins located on the arc-trench gap) is poorly understood, primarily because we lack detailed and published high resolution seismic sections that are calibrated by wells. In some cases, forearc basins may be dominated by gravity sliding and associated normal faulting; in others, their evolution may be influenced by the tectonic evolution of the underlying accretionary wedge; and, finally, some authors suspect an entirely extensional origin for these basins.

Backarc basins of the Western Pacific type as well as the "continental" backarc basins of the Pannonian type all appear to be initiated by a rifting phase and followed by later subsidence. A thermal origin associated with crustal stretching seems likely for these basins (McKenzie, 1978b; Molnar and Atwater, 1978; and Uyeda and Kanamori, 1979).

Subsidence of the Gulf of Lion - a backarc

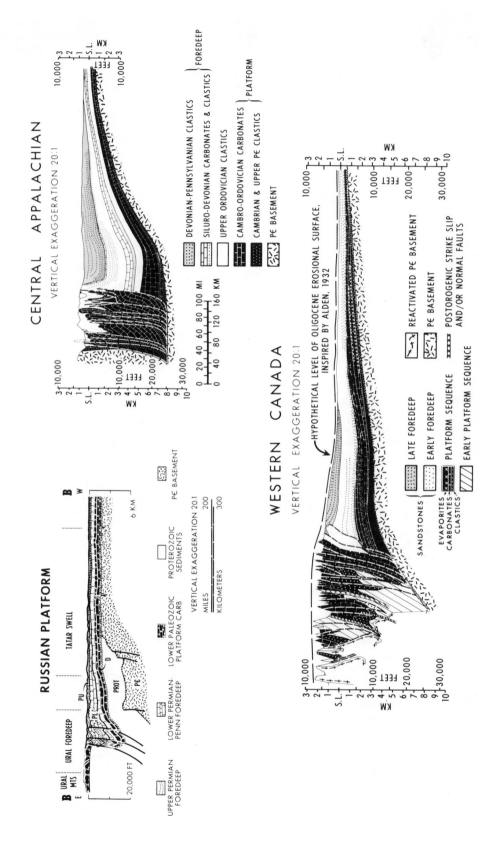

Fig. 7. Examples of foredeep basins from top to bottom. (a) Ural foredeep, (b) Appalachian foredeep, (c) Rocky Mountain foredeep (after Bally and Snelson, 1980, with permission of Can. Soc. Petrol. Geologists).

KNOXVILLE AREA VALLEY & RIDGE BELT GEOL. QUADRANGLES

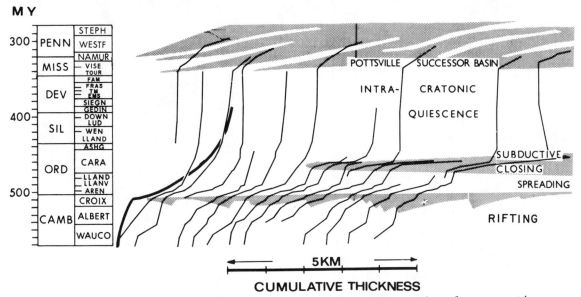

Fig. 8. Trinomial diagram illustrating cumulative sedimentation along a section across the Tennessee thrust belt, after data in published geological quadrangle maps. Vertical axis is in time; horizontal axis contains cumulative formation thickness in individual curves arranged palinspastically along a cross section. Each curve represents data taken from a quadrangle map. The thick line is the Sclater-Francheteau curve of oceanic subsidence for comparison. The deviation of the time-thickness plots from the Sclater-Francheteau curve reflects reflects the closing of the ocean during the Upper Ordovician. It is assumed that cumulative sedimentation reflects crustal subsidence except in the deep water sediments of the (Middle Ordovician) Tellico-Sevier shale complex. Shaded areas are times and spaces of crustal unrest, including a late Cambrian event of rifting, a Middle Ordovician and a Carboniferous foredeep event. (Figure and caption after Roeder et al., 1978, with permission of U. of Tennessee, Dept. of Geol. Sciences.)

basin - has been explained by Steckler and Watts (in preparation) in terms of thermal cooling, following substantial lithospheric stretching (i.e., by about a factor of 10). These authors note that geological evidence does not support such large amounts of stretching of the continental crust; and, therefore, according to Steckler and Watts, heating associated with stretching has to be supplemented by additional heating mechanisms, such as pre-existing high thermal gradients or active heating due to deep seated mantle processes. In any event, their study shows that subsidence for backarc basins may be modeled using criteria and methods that are similar to those used for passive margins. It has, however, been noted that backarc basins tend to survive only for limited periods, say (60 Ma) while passive margins form and subside over much longer time spans. In other words, the initial genesis and subsidence history may be similar, but backarc basins appear to have a built-in self-destruct mechanism.

The Pannonian Basin has been studied in some

detail, and its crustal structure is rather well known (Horvath and Stegena, 1977).

Recently Sclater et al. (1980) have looked at the intra-Carpathian episutural basins and their thermal origin. For a number of peripheral basins (e.g., the Vienna Basin), these authors postulate a two-fold stretching during Early-Middle Miocene, leading to a very fast initial subsidence and slow later subsidence due to conductive cooling of the lithosphere. The more central basins of Hungary (e.g., Pannonian Basin) prove more difficult to explain. A model assuming stretching by a factor of three could explain the thermal subsidence and the heat flow. Unfortunately, this is not supported by the available geologic information, which suggests the absence of a well defined initial subsidence. Sclater et al. offer alternative solutions: either a two-fold stretching accompanied by subcrustal attenuation of the lithosphere or else attenuation of the whole subcrustal lithosphere and part of the crust by subcrustal melting and erosion.

Finally, there is a group of *basins related*

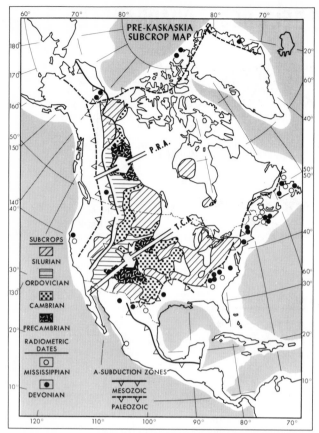

Fig. 9 (a). Pre-Kaskaskia subcrops. This map shows the generalized distribution of early Paleozoic beds under the unconformity at the base of the (Devonian-Mississippian) Kaskaskia sequence of Sloss (1972a). The Peach River Arc (P.R.A.) and the Transcontinental Arch (T.C.A.) represent major uplifts that occurred in pre-Mid Devonian and post-Silurian times. The apparent left lateral offset of the Precambrian basement in the area of the T.C.A. is due to later upper Paleozoic strike-slip faulting and related to the collision tectonics of the Marathon-Ouachita foreland.

to the genesis of major shear systems (e.g., San Andreas fault related basins of California, Basin and Range type basins, etc.). Some of these basins clearly owe their origin to crustal stretching, while others have to be viewed in a compressional context. Turcotte and McAdoo (1979) have made the case for thermally induced subsidence in the Los Angeles basin. In general, however, - as with other basin types - we still lack adequately precise subsurface data (particularly deep seismic reflection data) that would allow a realistic quantitative verification of any hypothesis.

The hazards of classification, such as the one adopted in this summary (Table 1), are that it is often difficult to find the right pigeon-

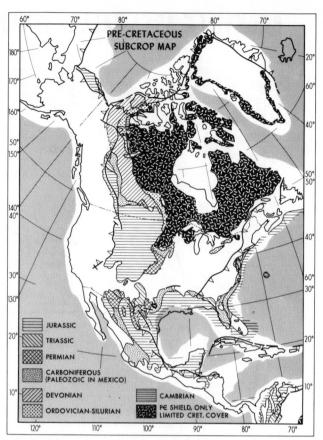

Fig. 9 (b). Pre-Cretaceous subcrops of North America. These subcrops reveal significant westerly differential tilting movements that precede the deposition of the Cretaceous fore-deep clastic wedge of the Western Interior.

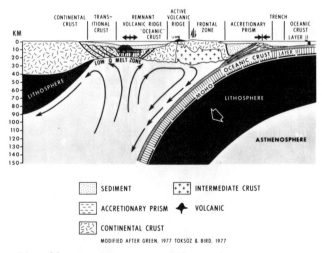

Fig. 10. Active margin and the formation of episutural sedimentary basins (modified after Green, 1977, and Toksöz and Bird, 1977).

hole for certain basins. An example of this is the class "inland seas" that is used by Artyush-kov, which includes the S. Caspian Basin, the Black Sea, the Western Mediterranean and others. Although some of these basins can be interpreted as backarc basins, this cannot reasonably be done for the S. Caspian Basin; and the interpre-tation of the Black Sea as a preserved Creta-ceous backarc basin (Letouzey et al., 1977) is highly speculative. For both basins, Artyushkov has proposed an origin due to the eclogitization of the lower crust, a process which could also be invoked for a number of other backarc basins. Artyushkov's mechanism differs from the "basifi-cation" or "oceanization" mechanism proposed by Beloussov (1968), because "oceanization" involves a change in the composition of the crust that is caused by a supply of ultrabasic material from the mantle.

Concerning Global Simultaneity of Subsidence and Uplift Episodes

A discussion on subsidence would not be com-plete without a short discussion of the world-wide correlatability of subsidence and uplift episodes. A new perspective in this area has been provided by Vail et al. (1977). These authors have developed a methodology that per-mits us to recognize unconformity-bounded sequences on seismic reflection lines. They offer a scheme to explain global correlations of seismostratigraphic sequences in terms of global eustatic sea level changes (Fig. 11). Vail et al., are very careful and correctly emphasize that eustatic sea level changes do not neces-sarily correlate with transgressions and regressions, because the latter indicate only the balance between the rate of sea-floor sub-sidence and the rate of sediment supply. The curves of Vail et al., show slow sea level rises and abrupt sea level falls, a picture

which appears to contrast with the concept often expressed by geologists, that transgres-sions are relatively quick and short-lived events. Such transgressions are likely to reflect rate changes in eustatic oscillations where for a brief period the sediment supply is unable to keep up with the subsidence.

Donovan and Jones (1979) pointed out that the curves published by Vail et al., are based on seismic and borehole data that are not pub-lished. Therefore, the curves officially remain uncalibrated. One may suspect that the horizontal "shoulders" of the major cycles may not be very crisply defined, because in reality the paleontologic age bracket of a typical hia-tus inferred from a reflection seismic line is in most cases rather loosely defined.

Donovan and Jones (1979) also have reviewed some of the causes for worldwide sea level changes and determined that typically changes in the volume of land ice led to sea level changes of about 150 m (or 1 cm/year), changes in the volume of ocean ridges could lead to sea level changes of about 300 m (1 cm/1,000 years), and the dessication of isolated ocean basins could lead to sea level changes of about 15 m (1 cm/year).

Sleep (1976) and Keen (1979) have evaluated the effects of sea level changes on subsidence. The effect on subsidence is minimal, but as Sleep points out, small sea level fluctuations may well produce significant unconformities even in rapidly subsiding basins.

All this suggests that sea level changes have relatively little impact on subsidence. In fact, Vail et al.'s cycles may correspond more to wide-spread and correlatable subsidence and uplift episodes. In this view, eustatic sea level changes may be subordinate to worldwide tectonic cycles, and the major remaining question is: Are intraplate subsidence and uplift episodes to some degree correlatable on a worldwide scale? Sloss (1972a, b and 1978), Rona (1973), Whitten (1976) and Soares et al. (1978) all suggest that this may be the case.

Gilluly (1973) emphasized the steady nature of plate motions, but Dewey (1975) in a very lucid account described how global changes in slip rates may lead to the large scale reorgani-zation of plate boundary configurations that would be expressed in the geologic record as global tectonic episodicity. In this perspec-tive, the "normal" part of Vail et al.'s curves (see Fig. 11) is the one interpreted as slow rise of the sea level, whereas the flat shoul-ders interpreted as a sudden fall of sea level constitute the "abnormal" event.

It stands to reason that the "normal" por-tion of the curves represent relatively steady "plate tectonics as usual processes" that are responsible for mountain tectogenesis. This view coincides with Johnson's (1971) observation that the rate of subsidence of cratonic basins correlates well with periods of mountain build-

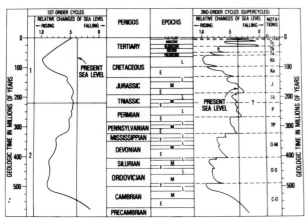

Fig. 11. First and second order global cycles of relative change of sea level during Phanero-zoic times (after Vail et al., 1977).

ing. On the other hand, the shoulders of Vail et al.'s curves would mark the slowdown and cessation of one major plate tectonic regime and its configuration and the beginning of another new plate tectonic configuration. During one such configuration, different modes of tectogenesis occur simultaneously in different parts of the world (see, also, Sloss, 1980).

It is concluded that the global cycles published by Vail et al. and the sequences described by Sloss and his followers may indicate widespread (but not globally ubiquitous) correlation of phases of basin subsidence. Unconformities that separate the cycles are the response to short periods of plate reorganization (for instance, due to continental collisions, ridge-continent collisions, changes in spreading systems, collapse of large marginal sea domains, etc.) that are followed by new subsidence cycles.

The "global" unconformities that separate major stratigraphic sequences also correspond rather well to Stille's (1924) orogenic phases, and recently Schwan (1980) considered the correlatability of these Alpino-type orogenies with discontinuities in sea-floor spreading (Fig. 12). Ziegler (1978) showed similar relations, but in his case he ties them also to unconformities that occur in the cratonic realm of Northwestern Europe.

In the light of Gilluly's eloquent criticism, it would be more plausible to relate roughly correlatable "worldwide" (better: widespread, because they are not ubiquitous) unconformities to major plate reorganizations, which are not necessarily orogenic or mountain building tectonic events. Such unconformities simply reflect uplifts which may affect cratonic and non-cratonic realms in different ways. With resumption of "plate tectonics as usual" after a major global reorganization, subduction-related tectogenesis and subsidence in cratonic realms would resume.

Whether the synchroneity of globally widespread phases of basin subsidence and uplift phases is controlled by the rhythm of asthenospheric flows between continents and oceans (Sloss and Speed, 1974; Sloss, 1980) has yet to be supported by geophysical observations.

All our arguments are not yet sufficiently precise to come up with a systematic review of worldwide basin-forming events, and thus one is limited to expressing only a few guesses:
1. The papers in this volume and other publications dimly suggest that major worldwide rifting events and the formation of associated graben systems occurred during the latest Precambrian. In a plate tectonic perspective, this could indicate the breakup of a late Precambrian supercontinent which, so far, is also only vaguely circumscribed by paleomagnetic data (Pangea E of Morel and Irving, 1978). The relation of such an event to widespread Panafrican-Baikalian orogenic activities also remains obscure.
2. The breakup of a supercontinent like Pangea since Triassic times proceeds in major segments, so that the inception and duration of rifting may differ for the different segments. The subsidence due to cooling associated with the opening of segments of an ocean begins at different times during the dispersal of Pangea, i.e., Mid-Jurassic for the North Atlantic, Lower Cretaceous for the South Atlantic, late Cretaceous for the Greenland-Northern Europe Atlantic, Eocene for Australia-Antarctica, upper Miocene for the Red Sea and Gulf of Aden, etc.
3. Accepting plate tectonics as a working hypothesis, it would follow that the subsidence history of foredeeps should correlate with the subsidence history of that segment of a spreading ocean that is responsible for the subduction and mountain building process which has to "absorb" the effects of ocean spreading.

Coney (1979) has illustrated this point by relating the opening of the Atlantic to the Mesozoic-Cenozoic evolution of the Western Cordillera. The over-all synchroneity of the subsidence on the Atlantic margin and the Mesozoic-Paleogene foredeep of North America support these relations.

Similarly, the opening of the Indian Ocean and the subsidence history of its margins may be tied to the subsidence history of foredeeps of the Alpine-Himalayan system.
4. The majority of the backarc basins of the world appear to be initiated by thermal rifting and/or stretching processes during late Paleocene times (although there are a number of sig-

PERIODS	EPOCHS	AGE IN M.Y.	DISCONTINUITIES OF OCEAN – FLOOR SPREADING IN THE NORTH ATLANTIC (i.a. PITMAN AND TALWANI 1972; DEWEY ET AL 1973; VOGT 1974, 1975)	UNCONFORMITIES OF OROGENIC PHASES IN EUROPE AND NORTH AMERICA (i.a. STILLE 1924; KRÖMMELBEIN 1977; TOLLMANN 1966; TRÜMPY 1973)
TERTIARY	PLIOCENE	5	10 – 9 M.Y.	STYRIAN
	MIOCENE		~ 17 M.Y.	DO.
	OLIGOCENE	23		
		36	42 – 38 M.Y.	ILLYIAN (1. PYRENEAN)
	EOCENE			
	PALEOCENE	53.5	53 M.Y.	LARAMIDE
		65	63 M.Y.	DO.
CRETACEOUS	UPPER CRETACEOUS		80 – 75 M.Y.	SUBHERCYNIAN
		100		
	LOWER CRETACEOUS		115 – 110 M.Y.	AUSTROALPINE
		136		
JURASSIC	UPPER JURASSIC	148	148 M.Y.	NEVADAN (1. LATE CIMMERIAN)
		157		

AFTER SCHWAN, 1980

Fig. 12. Correlation of discontinuities in ocean-floor spreading and unconformities of orogenic phases (after Schwan, 1980).

nificant exceptions). Further subsidence possibly due to cooling and/or spreading in the marginal basins occurs during the Neogene. Rona and Richardson (1978) suggest that during the Eocene, a major reorganization of global plate motion pattern occurred where motions with large N-S components were replaced by motions with large E-W components because of an increase in length in collisional boundaries. Consequently, increased E-W sea-floor spreading that appears to be associated with a number of backarc basins may be induced by such reorientations.

The above-mentioned speculations provide possible rationales for an over-all correlatability of worldwide subsidence episodes of differing character that are based on the plate tectonic hypothesis. Such rationales become much more tenuous when applied to cratonic basins (syneclises) and arches (anteclises). Of course, the separation of cratonic basins and foredeeps is somewhat arbitrary, and authors differ on their definitions. But the major problem is that we need reasonably accurate cross sections based on reflection seismic data and quantitative subsidence reconstructions across many cratonic basins in the world before we undertake any correlation. These are not currently available.

Adherents of fixist concepts (Beloussov, 1962, 1975) or of global expansion (Carey, 1977) would search for different rationales to explain any global correlatability of subsidence, but these alternative hypotheses are not nearly so dependent on establishing such global correlations. For instance, the mechanisms proposed in Artyushkov's contribution in this volume are interesting in that they accept plate tectonics but assume the existence of long-lasting plumes that channel hot differentiates from the lower mantle, which may migrate in traps that are formed at the base of the lithosphere. As such traps may exist for long periods, they may well explain repeated uplift in certain areas.

Future Research

Our speculations on global correlatability lead naturally to questions that may concern future research on basin subsidence.
1. It is obvious from our summary that we lack precise basinwide subsidence measurements for many areas. Also, the precision and documentation of available measurements leaves much to be desired. Van Hinte (1978) has laid out the procedure for a more reliable analysis of basin subsidence, which should include (a) a proper evaluation of paleontologic and paleoecologic information, (b) compaction corrections, and (c) an indication of the origin and the radiometric calibration of the time scale used. There is still substantial uncertainty in the choice of radiometric scales (Cohee et al., 1978), and this is particularly critical for subsidence studies of Paleozoic and late Precambrian sequences.

The methodology for the application of geophysical model studies to subsidence studies is contained in the recently published papers by Steckler and Watts (1978), Turcotte and McAdoo (1979), Keen (1979), Royden et al. (1980), and Sclater and Christie (1980). A number of these studies also indicate their practical importance for the evaluation of the petroleum potential of sedimentary basins. Clearly, the maturation history of organic-rich petroleum source beds is influenced by the thermal evolution of sedimentary basins.
2. Subsidence studies that are based on well data only and that are not complemented by reflection seismic lines are inherently weakened by the circumstance that most hydrocarbon exploration wells are designed to test anomalies. To understand the size, the local or regional significance of such anomalies, it is desirable whenever possible to complement subsidence data with reflection seismic sections that hopefully would reveal more about the nature of the bottom of the basin (i.e., extensive faulting may suggest an initial thermal event, or, conversely, lack of faulting would favor models emphasizing flexural bending).
3. Causes for subsidence and major uplifts within basins are often postulated to be located in the lower lithosphere (lower crust and mantle) and the asthenosphere. For this reason, geophysical studies (reflection, refraction and wide angle reflection surveys) of the lower crust the mantle, and the asthenosphere are most important. Cooperation with surface and subsurface geologists would insure that these studies are laid out to adequately characterize the geologic dimensions of basins and the uplifts that form their margins (i.e., deep seismic sounding experiments - refraction, wide angle reflection, and deep crustal reflection, etc. - should cover both the basin and the adjacent flanks and uplifts). With the notable exception of the USSR and parts of eastern Europe, deep lithospheric and asthenospheric data obtained with the intent to characterize areas of subsidence and adjacent areas of uplift are rarely available. Yet such data are needed to enable us to differentiate among various proposed subsidence mechanisms.
4. A significant counterpart of basin subsidence is related to uplifts that form its rims. The genesis of such uplifts is poorly known and the incentive to study them is limited, because erosion has eliminated much of the record. However, it must be emphasized that regional and supraregional subcrops such as those described by Levorsen (1960) and Cook and Bally (1975) and Fig. 9a and b provide valuable information concerning tilting movements. Together with deep geophysical information, such studies could lead to the formulation of alternative models for the formation of arches.
5. A thermal origin of many basin types is postulated by a number of geophysical models. To

test them, heat flow studies of the type discussed by a number of our authors are needed.
It is most desirable to complement such work
with studies that attempt to unravel the thermal
evolution of the basin fill. Such studies
include clay compaction, the metamorphism of
clay minerals, organic maturation studies based
on various indicators (coalification, vitrinite
reflectance, spore translucency, conodonts,
hydrocarbon maturity, etc.).

These suggestions for future research should
be amplified by a plea to continue traditional
studies (i.e., stratigraphy and all other geologic compilation directed toward quantitative
knowledge of subsidence history, gravity, mag-
netics, etc.). But above all, we would like to
stress that subsidence and basin evolution studies are most successful when done by teams of
various specialists; no single individual is
able to master all the methods that are required
to study the evolution of basins. In a nut-
shell, multidisciplinary geological, geochemical, paleontological, and geophysical work is a
"must" for subsidence studies.

Acknowledgements: I thank Shell Oil Company
for providing time and help to edit the papers
of this part and for permission to publish this
review. I would particularly like to thank
H. Scott and D. Branson for preparing the illus-
trations and K. Ziegler and J. Cartwright for
the retyping of all manuscripts of Part I of this
volume.

References

Bally, A. W., Canada's passive continental margins - a review, Mar. Geophys. Res., 2(4),
327-340, 1976.

Bally, A. W., and S. Snelson, Realms of subsidence, Can. Soc. Petrol. Geol. Mem., 6, 1-94,
1980.

Bally, A. W. et al., A sketch of the tectonics of
China, Open File Rep., 80-501, U.S. Geol.
Surv., Reston, VA, 1980.

Beaumont, C., The evolution of sedimentary
basins on a viscoelastic lithosphere: theory
and examples, Geophys. J. Roy. Astron. Soc.,
55, 471-498, 1978.

Beaumont, C., Foreland Basins, submitted to
Geophys. J. Roy. Astron. Soc., 1980.

Beaumont, C., and J. F. Sweeney, Graben generation of major sedimentary basins, Tectonophysics, 50, T19-T23, 1978.

Beaumont, C., On rheological zonation of the
lithosphere during flexure, Tectonophysics,
59, 347-365, 1979.

Beloussov, V. V., Basic Problems in Geotectonics, 809 pp., McGraw Hill, New York, 1962.

Beloussov, V. V., The crust and upper mantle of
the oceans (in Russian), M., Nauka, 1968.

Beloussov, V. V., Foundations of geotectonics
(in Russian), Moscow, Nyedra, 260 pp., 1975.

Bird, Peter, Initiation of intracontinental
subduction in the Himalaya, J. Geophys. Res.,
83(B10), 4975-4987, 1978.

Bird, Peter, Continental delamination and the
Colorado Plateau, J. Geophys. Res., 84, 7561-
7571, 1979.

Brewer, J. A., S. B. Smithson, J. E. Oliver,
S. Kaufman, and L. D. Brown, The Laramide
Orogeny: Evidence from COCORP Deep Crustal
Seismic Profiles in the Wind River Mountains,
Wyoming, Tectonophysics, 62, 165-187, 1980.

Briggs, G., and D. Roeder, Sedimentation and
plate tectonics, Ouachita Mountains and
Arkoma Basin, in A Guidebook to the Sedimen-
toloty of Paleozoic Flysch and Associated
Deposits, Ouachita Mountains-Arkoma Basin,
Oklahoma, Dallas Geol. Soc., 1-22, 1975.

Caldwell, J. G., W. F. Haxby, D. E. Karig, and
D. L. Turcotte, On the applicability of a
universal elastic trench profile, Earth
Planet. Sci. Lett., 31, 239-246, 1976.

Caldwell, J. G., and D. L. Turcotte, Dependence
of the thickness of the elastic oceanic
lithosphere on age, J. Geophys. Res., 84,
7552-7576, 1979.

Carey, S. W., The expanding earth, Developments
in Geotectonics Ser., No. 10, 488 pp., Amster-
dam, Elsevier, 1977.

Cohee, V. G., M. F. Glaessner, and H. G. Hed-
berg, eds., Contributions to the geologic
time scale, Am. Assoc. Petrol. Geol. Studies
in Geology, No. 6, 388 pp., 1978.

Coney, P. J., Mesozoic-Cenozoic Cordilleran
plate tectonics, Geol. Soc. Amer. Mem.,
152, 33-50, 1979.

Cook, T. D., and A. W. Bally, eds., Stratigraphic Atlas of North and Central America,
272 pp., Princeton University Press, 1975.

Dewey, J. F., Finite plate implications: Some
implications for the evolution of rock masses
at plate margins, Amer. J. Sci., 275-A, 260-
284, 1975.

Donovan, D. T., and E. J. W. Jones, Causes of
worldwide changes in sea level, J. Geol. Soc.
Lond., 136, 187-192, 1979.

Fischer, A. G., Geological time distance rates:
the Bubnoff unit, Geol. Soc. Amer. Bull., 80,
549-552, 1969.

Gilluly, J., Steady plate motion and episodic
orogeny and magmatism, Geol. Soc. Amer.
Bull., 84, 499-514, 1973.

Horvath, F,. and L. Stegena, The Pannonian Basin:
a Mediterranean interarc basin, in Structural
History of the Mediterranean Basins, (B. Biju-
Duval and L. Montadert, eds.), Editions Technip, 133-142, 1977.

Johnson, J. G., Timing and coordination of orogenic, epirogenic and eustatic events, Geol.
Soc. Amer. Bull., 82, 3263-3298, 1971.

Keen, C. E., Thermal history and subsidence of
rifted continental margins - evidence from
wells on the Nova Scotian and Labrador
shelves, Can. J. Earth Sci., 505-522, 1979.

LePichon, X., and J. C. Sibuet, Passive margins: a model of formation, manuscript submitted to J. Geophys. Res., 1980.

Letouzey, B., B. Biju-Duval, A. Dormel, R. Gonnard, K. Kritschev, L. Montadert, and D. Sungurly, The Black Sea: a marginal basin geophysical and geological data, in Structural History of the Mediterranean Basins, (B. Biju-Duval and L. Montadert, eds.), Editions Technip, 363-376, 1977.

Levorsen, A. I., Paleogeologic Maps, W. H. Freeman and Company, 1960.

Lyustikh, E. N., On the thalassogenesis hypotheses and on crustal blocks (in Russian), Izv. AN SSSR, Ser. Geophys., 11, 1959.

Matthews III, V., ed., Laramide folding associated with basement block faulting in the western United States, Geol. Soc. Amer. Mem., 152, 370 pp., 1978.

McKenzie, D., Some remarks on the development of sedimentary basins, Earth Planet. Sci. Lett., 40, 25-32, 1978.

Molnar, P., and T. Atwater, Interarc spreading and cordilleran tectonics as alternates related to the age of subducted oceanic lithosphere, Earth Planet. Sci. Lett., 41, 330-340, 1978.

Montadert, L., D. G. Roberts, O. de Charpal, and P. Guennoc, Rifting and subsidence of the northern continental margin of the Bay of Biscay, Init. Repts. of the Deep Sea Drilling Proj., XLVIII, 1025-1060, 1979.

Morel, P., and E. Irving, Tentative paleocontinental maps for the early Phanerozoic and Proterozoic, J. Geol., 86, 535-561, 1978.

Neugebauer, H. J., and T. Spohn, Late stage development of mature Atlantic-type continental margins, Tectonophysics, 50, 275-305, 1978.

Parsons, B., and P. Molnar, The origin of outer topographic rises associated with trenches, Geophys. J. Roy. Astron. Soc., 45, 707-712, 1976.

Petters, S. W., West African cratonic stratigraphic sequences, Geology, 7, 528-531, 1979.

Roeder, D., Geodynamics of the Alpine-Mediterranean System - A Synthesis, Eclogae Geol. Helv., 2, 1980.

Roeder, E., D. E. Gilbert, Jr., and W. D. Witherspoon, Evolution and macroscopic structure of Valley and Ridge thrust belt, Tennessee and Virginia, 25 pp., Univ. Tenn. Geol. Sci. Studies in Geology, 2, Knoxville, 1978.

Rona, P. A., Worldwide unconformities in marine sediments related to eustatic changes of sea level, Nature Phys. Sci., 244, 25-26, 1973.

Rona, P. A., and E. S. Richardson, Early Cenozoic global plate reorganization, Earth Planet. Sci. Lett., 40, 1-11, 1978.

Royden, L., J. G. Sclater, and R. P. von Herzen, Continental margin subsidence and heat flow: important parameters in formation of petroleum hydrocarbons, Amer. Assoc. Petrol. Geol. Bull., 64, 173-187, 1980.

Schwan, W., Geodynamic peaks in alpino-type orogenies and changes in ocean-floor spreading during late Jurassic-late Tertiary time, Amer. Assoc. Petrol. Geol. Bull., 64, 359-373, 1980.

Sclater, J. G., and P. A. F. Christie, Continental stretching: An explanation of the post mid-Cretaceous subsidence of the Central North Sea Basin, in press, 1980.

Sclater, J. G., L. Royden, F. Horvath, B. C. Burchfiel, S. Semken, and L. Stegena, The formation of the intra-Carpathian basins as determined from subsidence data, manuscript, 1980.

Sclater, J. G., W. Royden, S. Sempken, C. Burchfield, L. Stegena, and F. Horvath, Continental stretching: An explanation of the late Tertiary subsidence of the Pannonian Basin, in preparation, 1980.

Sclater, J. G., L. Royden, and F. Horvath, The formation of the intra-Carpathian basins determined from subsidence history, Abstr., Extensional Tectonics Associated with Convergent Plate Boundaries, Roy. Soc. London, 1980.

Sleep, N. H., Platform subsidence mechanisms and "eustatic" sea-level changes, Tectonophysics, 36, 45-56, 1976.

Sleep, N. H., J. A. Nunn, and Lei Chou, Platform basins, 624 pp., Ann. Rev. Earth and Planet. Sci., 8, 1980.

Sloss, L. L., Synchrony of Phanerozoic sedimentary tectonic events of the North American craton and the Russian platform, Twenty Fourth Session, International Geological Congress, Section 4, 24-32, 1972a.

Sloss, L. L., Concurrent subsidence of widely separated cratonic basins, Geol. Soc. Amer. Program Abs., 4, 668-669 (Fide Whitten, 1976), 1972b.

Sloss, L. L., Global sea level change: A review from the craton, Amer. Assoc. Petrol. Geol. Mem., 29, 461-467, 1978.

Sloss, L. L., Subsidence of continental margins: The case for alternatives to thermal contraction, Inter-Union Commission on Geodynamics, Working Group 8, Final Report, R. A. Scrutton, ed., in preparation, 1980.

Sloss, L. L., and R. C. Speed, Relationships of cratonic and continental margin tectonic episodes, in Tectonics and Sedimentation, (N. R. Dickinson, ed.), Soc. Econ. Paleont. Mineral. Spec. Pub. 22, 98-119, 1974.

Smithson, S. B., J. A. Brewer, S. Kaufman, J. E. Oliver, and C. Hurich, Nature of the Wind River Thrust, Wyoming, from COCORP deep reflection and gravity data, Geology, 6, 648-652, 1978.

Smithson, S. B., J. A. Brewer, S. Kaufman, and J. E. Oliver, Structure of the Laramide Wind River Uplift, Wyoming, from COCORP deep reflection data and from gravity data, J. Geophys. Res., 84(B11), 5955-5972, 1979.

Soares, P. C., P. M. B. Landim, and V. J. Fulfard, Tectonic cycles and sedimentary sequences in the Brazilian intracratonic basins, <u>Geol. Soc. Amer. Bull.</u>, <u>89</u>, 181-191, 1978.

Steckler, M. S., and A. B. Watts, Subsidence of the Atlantic-type continental margin off New York, <u>Earth Planet. Sci. Lett.</u>, <u>41</u>, 1-31, 1978.

Steckler, M. S., and A. B. Watts, The Gulf of Lion: Subsidence of a young continental margin. Submitted to <u>Nature</u>, 1980.

Stille, H., Grundfragen der vergleichenden Tektonik, Berlin, 1924.

Turcotte, D. L., and D. C. McAdoo, Thermal subsidence and petroleum generation in the southwestern block of the Los Angeles basin, <u>J. Geophys. Res.</u>, <u>84</u>, 3460-3464, 1979.

Uyeda, S., and H. Kanamori, Back-arc opening and the mode of subduction, <u>J. Geophys. Res.</u>, <u>83</u>, 1049-1061, 1979.

Vail, P. R., R. M. Mitchum, Jr., R. G. Todd, J. M. Widmier, S. Thompson III, J. B. Sangree, J. N. Bubb, and W. G. Hatlelid, Seismic stratigraphy and global changes of sea level, in <u>Seismic Stratigraphy - Applications to Hydrocarbon Exploration</u>, C. E. Payton, ed.,

Amer. Assoc. Petrol. Geol. Mem., <u>26</u>, 49-212, 1977.

Van Hinte, J. E., Geohistory analysis - applications of micropaleontology in exploration geology, <u>Amer. Assoc. Petrol. Geol. Bull.</u>, <u>62(2)</u>, 201-222, 1978.

Vetter. U. R., and R. O. Meissner, Rheologic properties of the lithosphere and applications to continental margins, <u>Tectonophysics</u>, <u>59</u>, 367-380, 1979.

Watts, A. B., and M. Talwani, Gravity anomalies seaward of deep sea trenches and their tectonic implications, <u>Geophys. J. Roy. Astron. Soc.</u>, <u>36</u>, 57-90, 1974.

Watts, A. B., and M. S. Steckler, Subsidence and eustasy at the continental margin of eastern North America, <u>Maurice Ewing Symposium, Series 3</u>, <u>Amer. Geophys. Union</u>, 218-234, 1979.

Whitten, E. T. H., Cretaceous phases of rapid sediment accumulation continental shelf, eastern U.S.A., <u>Geology</u>, <u>4(3)</u>, 237-240, 1976.

Ziegler, P. A., Northwestern Europe: Tectonics and Basin Development, <u>Geologie en Mijndouw</u>, 57, 589-626, 1978.

MODELS FOR THE EVOLUTION OF SEDIMENTARY BASINS

D. L. Turcotte

Department of Geological Sciences, Cornell University, Ithaca, New York 14853

Abstract. Two mechanisms for the formation of sedimentary basins are well understood. The first is the deposition of sediments in a topographic low. The isostatic subsidence causes the total depth of sediments to be two to three times the initial topographic anomaly. These types of basins occur at continental margins and in areas of continental rifting. A second cause of sedimentary basins is thermal subsidence. As the lithosphere cools, its density increases and it subsides due to isostasy. If the subsiding basement is covered with sediments, a sedimentary basin is formed. Phase changes in the crust may also lead to subsidence. Tectonic processes can cause crustal thinning and sediments can fill the resulting topographic low. For sedimentary basins with horizontal dimensions less than a few hundred kilometers the flexural rigidity of the lithosphere can restrict subsidence. However, differential vertical displacements can take place on faults.

Introduction

The simplest model for the development of sedimentary basins is to fill topographic lows with sediments. Subsidence will occur due to the loading of the sediments so that the depth of the sedimentary basin will be greater than the initial depth of the topographic low. Topographic lows may be associated with oceanic crust. One example of sedimentary fill on oceanic crust would be major river deltas such as the Amazon and Niger. In other cases the oceanic crust may be of a more limited extent such as in the Gulf of California where the Colorado River provided sedimentary fill. In this case the basic mechanism for the creation of oceanic crust at a spreading center is providing the topographic low. In some cases topographic lows occur on continental crust when a region is surrounded by mountain ranges or crustal domes.

Subsidence can also result in the formation of sedimentary basins. The best documented cause of subsidence is the cooling and subsequent thermal contraction of the lithosphere. An example is the subsidence of the oceanic lithosphere as it ages to form the ocean basins. The oceanic lithosphere is thin where it is created at ocean ridges. As heat is conducted to the sea floor, the lithosphere cools and thickens. The lithospheric rocks become more dense and the ocean floor subsides according to the requirements of isostasy. It has been shown by Sleep (1971) that many sedimentary basins have exhibited thermal subsidence.

Alternative mechanisms for subsidence include crustal thinning and phase changes. If the continental crust is thinned, subsidence will result. Crustal thinning may occur due to tensional failure of the continental lithosphere or the gravitational forces present at continental margins. The most likely phase change that may result in subsidence is the transition of crustal basalt to eclogite.

Subsidence may be inhibited by the bending rigidity of the lithosphere. The ability of the earth's lithosphere to inhibit subsidence is determined by the magnitude of the flexural parameter. For typical continental lithosphere the flexural parameter has a value of about 80 km. Sedimentary basins with horizontal dimensions much larger than this value can subside isostatically. Subsidence of smaller sedimentary basins can be inhibited by the flexural strength of the lithosphere. An example is the Michigan basin. Small sedimentary basins or sections of sedimentary basins may subside isostatically if they are bounded by active faults. In some cases active faulting may play an important role early in the evolution of a sedimentary basin when the rates of subsidence are large. Subsequently, when the subsidence slows, the faults may lock and flexure plays an important role. A transition of this type appears to have occurred in the North Sea basin.

Isostasy

It has long been known that long wave length topography on the earth is in hydrostatic equilibrium. With a flat earth approximation this condition requires that

$$\int_0^{z_L} \Delta\rho\,dz = \text{const} \qquad (1)$$

The density (or density anomaly) $\Delta\rho$ integrated over a depth z_L is a constant. This is known as the principle of isostasy and the depth z_L is known as the depth of compensation.

With plate tectonics it is now recognized that the near surface rocks which make up the plates behave as an elastic solid over geological times, but beneath the plates the solid mantle rocks exhibit fluid behavior on geological time scales. That portion of the mantle as well as the crust that makes up the plates is known as the lithosphere. In the ocean basins the lithosphere has an average thickness of about 110 km and beneath stable continental regions a thickness of about 180 km. In considering the application of the principle of isostasy, it is appropriate to equate the depth of compensation in Equation (1) with the thickness of the lithosphere.

An elementary application of the principle of isostasy is the subsidence caused by sedimentary fill in a topographic low. This is illustrated in Figure 1a. The depth of the topographic low is d and it is assumed to be filled with water, density ρ_w. If the topographic low is filled with sediments of density ρ_s the depth of the sediments obtained from Equation (1) is

$$d_s = \left(\frac{\rho_m - \rho_w}{\rho_m - \rho_s}\right) d \qquad (2)$$

Taking $\rho_w = 1$ Mg/m^3 and $\rho_m = 3.4$ Mg/m^3, the dependence of d_s on d is given in Figure 1b for several values of ρ_s.

An example of the sedimentary fill of a topographic low would be the sedimentation associated with major river deltas at continental margins. A typical depth for the major ocean basins is 6 km. From Figure 1b it is seen that the depth of the sedimentary fill can be as great as 20 km due to the isostatic subsidence of the ocean basin in response to the sedimentary load.

Isostatic subsidence may also occur without sedimentary fill. The best documented cause of isostatic subsidence is the cooling and subsequent increase in density of the lithosphere. An example of thermal subsidence is the development of the deep ocean basins. The oceanic lithosphere is created from hot mantle rock at

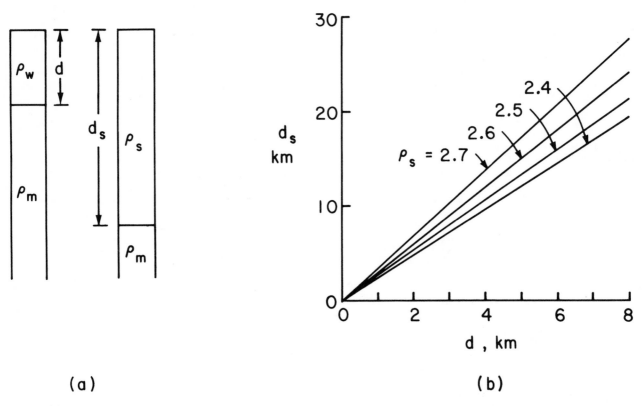

(a)   (b)

Fig. 1. (a) Depth of a water-filled basin d compared with the depth due to isostatic subsidence when the basin is filled with sediment. (b) Depth of a sediment-filled basin d_s as a function of the depth of a water-filled basin d for various sediment densities.

mid-ocean ridges. The loss of heat at the sea floor causes the oceanic lithosphere to thicken and cool. The thermal structure of the oceanic lithosphere can be approximated by the solution to the equation for the one-dimensional cooling of a semi-infinite half-space (Turcotte and Oxburgh, 1967)

$$\frac{\partial T}{\partial t} = \kappa \frac{\partial^2 T}{\partial z^2} \qquad (3)$$

where t is the age of the lithosphere and z is the depth. The necessary initial and boundary conditions are that $T = T_m$ at $t = 0$, $T = T_s$ at $z = 0$, and $T \to T_m$ as $z \to \infty$, where T_m is the temperature of hot mantle rock and T_s is the temperature of the sea floor. The solution of Equation (3) that satisfies these boundary conditions is

$$\frac{T - T_s}{T_m - T_s} = \mathrm{erf}\left[\frac{z}{2}\left(\frac{1}{\kappa t}\right)^{1/2}\right] \qquad (4)$$

where erf is the error function.

The depth of the sea floor as a function of age can be obtained by applying the principle of isostasy, Equation (1), to the cooling oceanic lithosphere. The subsidence w of the sea floor relative to the ocean ridge where it is formed is related to the density distribution in the oceanic lithosphere by

$$(\rho_m - \rho_w)w = \int_0^\infty (\rho - \rho_m)\,dz \qquad (5)$$

where ρ_m is the density of the hot mantle rock and ρ_w the density of the sea water. As the oceanic lithosphere cools it becomes more dense by thermal contraction and sinks. The density in the oceanic lithosphere is related to the temperature by

$$\rho - \rho_m = \rho_m \alpha(T - T_m) \qquad (6)$$

where α is the volume coefficient of thermal expansion.

Substitution of Equations (4) and (6) into Equation (5) and integrating gives

$$w = \frac{2\rho_m \alpha}{(\rho_m - \rho_w)}\left(\frac{\kappa t}{\pi}\right)^{1/2}(T_m - T_s) \qquad (7)$$

The observed increase in depth away from the Mid-Atlantic Ridge and East Pacific Rise is given as a function of sea-floor age in Figure 2 (Parsons and Sclater, 1977). The observations are compared with the predicted subsidence given by Equation (7) with $\rho_m = 3.4$ Mg/m^3, $\alpha = 3 \times 10^{-5}$ °C^{-1}, $\kappa = 1$ μm^2/sec, and $T_m - T_s = 1300$°C. It is seen that the agreement

between theory and observations is quite good.

Some departure of the observed subsidence from the simple lithospheric cooling model occurs at ages greater than about 80 Ma. The ocean floor does not continue to subside as much as predicted by Equation (6). This is probably caused by an input of heat at the base of the lithosphere. Possible causes of this heating are mantle convection beneath the rigid lithosphere or frictional heating. An empirical equation which predicts the flattening of the subsidence curve with time is

$$w = w_r + w_o\left[1 - e^{-(t/\tau)^2}\right]^{1/4} \qquad (8)$$

For small times this reduces to (7) if

$$\tau = \frac{\pi}{\kappa}\left[\frac{w_o(\rho_m - \rho_w)}{2\rho_m \alpha (T_m - T_s)}\right]^2 \qquad (9)$$

Taking $w_o = 4$ km and the parameters as given above good agreement with observations is obtained as shown in Figure 2.

It was suggested by Sleep (1971) that thermal subsidence has occurred in many sedimentary basins. His results for the ratio of the depth of a sedimentary layer deposited at time t after the initiation of subsidence to the depth of basement are given in Figure 3 for the Michigan, Appalachian, Gulf Coast, and Atlantic Coast basins of the United States. Each data point is the result of averaging a large number of wells. Applying the subsidence equation given in Equation (8) the relative depth of the sedimentary layers is given by

$$\frac{w}{w_m} = 1 - \left[1 - e^{-(t/\tau)^2}\right]^{1/4} \qquad (10)$$

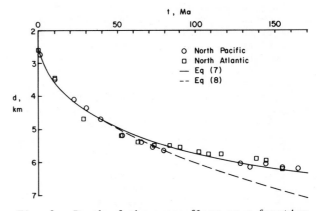

Fig. 2. Depth of the ocean floor as a function of age. The data points are the depths in the North Pacific and North Atlantic as given by Parsons and Sclater (1977). The dashed line is the cooling half-space model, Equation (7), and the solid line, the empirical correlation, Equation (8).

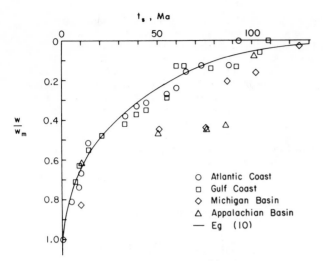

Fig. 3. The ratio of the depth w of a sedimentary layer deposited at time t_s to the depth of basement w_m as a function of the time t_s after the initiation of subsidence, for several sedimentary basins (after Sleep, 1971). The solid line is the theoretical subsidence from Equation (10) with $\tau = 80$ Ma.

where w is the depth to a sedimentary layer deposited at time t after the initiation of subsidence and w_m is the depth of the oldest sediments. Taking $\tau = 80$ Ma, good agreement is obtained with the data as is shown in Figure 3. The data points which lie below the curve can be attributed to periods when the basins were starved of sediments (Sleep, 1971). These results indicate that whatever mechanism is responsible for the subsidence of these basins, it has a thermal origin.

One mechanism for purely thermal subsidence would be an episode of crustal heating. The heating would result in uplift and erosion. At the end of the period of heating, the crust would cool by heat transfer to the surface. This would result in the type of subsidence curve illustrated in Figure 3.

A possible example of this type of subsidence would be passive continental margins. Presumably the early stages in the formation of an ocean resemble those seen today in the East African Rift. Extensive volcanism is occurring and there are regions of extensive crustal doming and erosion. The sedimentary basins on the continental margins of the eastern United States exhibit thermal subsidence as discussed above, and Turcotte et al.(1977) have suggested that this is the result of cooling after the opening of the Atlantic Ocean.

Not all basins which exhibit thermal cooling subsidence can be explained in this way. There is observational evidence that the Michigan basin experienced little or no uplift prior to the initiation of subsidence. Although the subsidence record of this basin suggests a thermal cooling history, a more complex origin must have been responsible for the initiation of subsidence.

Phase Changes

Isostasy requires that an increase in density of the rock in the crust or lithosphere will result in subsidence. One mechanism for increasing the density of rock is thermal contraction, but an alternative mechanism is for a solid-state phase change to occur. Although crustal rocks may undergo a wide range of phase changes, probably the most significant is the transformation of gabbro to the denser phase eclogite. A number of authors, for example, Joyner (1967), Collette (1968), Falvey (1974), and Haxby et al. (1976) have suggested that the gabbro-eclogite phase change caused basin subsidence. The importance of this phase change is that gabbro is a primary component of the lower continental crust; also the density change of about 0.5 Mg/m^3 (eclogite is the denser phase) is large.

One approach to determining the influence of phase changes on basin subsidence is to assume that the phase change is in thermodynamic equilibrium. As the pressure and temperature changes during subsidence, the phase change boundary will migrate so that it remains on the Clausius-Clapeyron curve. This problem has been treated by O'Connell and Wasserburg (1957) and by Mareschal and Gangi (1977).

However, this equilibrium approach is probably not appropriate for the gabbro-eclogite phase change. Laboratory studies (Green and Ringwood, 1972) have shown that gabbro is not stable at near-surface conditions even though it is the phase present in surface rocks. The conclusion is that the near-surface volcanic rocks are cooled sufficiently rapidly that the phase change gabbro to eclogite is quenched; the surface gabbros are in a metastable state. Therefore, the transformation of gabbro to eclogite is determined by a reaction rate. This rate process has been studied by Ahrens and Schubert (1975). If lower crustal rocks are heated, the reaction rate for the gabbro-eclogite phase change may increase sufficiently so that large volumes of rock may transform from gabbro to eclogite. The resulting increase in density will offset the decrease in density due to heating; subsequently thermal subsidence will occur when the rocks cool. Haxby et al. (1976) suggest that this is an important mechanism in the formation of the Michigan basin.

Crustal Thinning

On continents major areas of high topography are associated with areas of thick continental crust. The principle of isostasy is satisfied by having heavy mantle rock displaced by lighter crustal rock in order to compensate for the mass of the elevated topography. Variations in crustal thickness cause stress. Stress levels required to maintain the topography of mountains are of the order of 100 MPa (Artyushkov, 1973). Extreme variations in crustal thickness occur at continental margins. The resulting stresses may cause lower crustal material to flow from the thicker continental crust to the thinner oceanic crust. The result would be a thinning of the continental crust near the continental margin and a thickening of the adjacent oceanic crust (Bott and Dean, 1972). The resulting isostatic subsidence of the continental side of the continental margin could result in the formation of a sedimentary basin.

Another form of crustal thinning is associated with rift valleys. Rift valleys are believed to represent the failure mechanism of continental lithosphere under tension. Significant horizontal extension occurs in rift valleys. The near-surface expression is normal faulting and the formation of graben structures. These are the result of the brittle failure of the near-surface rocks. The deeper crustal rocks are likely to deform plastically. The horizontal extension of these rocks will lead to subsidence and a sedimentary basin in the rift valley. An example is the Rio Grande Rift in New Mexico.

Lithospheric Flexure

The basic hypothesis of plate tectonics is that the outer shell of the earth is rigid on geological time scales. Rigidity implies the ability to transmit elastic stresses. Vertical movements of the earth's surface can be inhibited by the bending rigidity of the near-surface rocks.

The surface plates are often referred to as the lithosphere. However, there are several definitions of the thickness of the lithosphere. The base of the lithosphere is probably best defined as the transition of a rigid rheology on geological time scales to a fluid rheology. This is likely to occur at a temperature of about 1200°C. This thermal lithosphere can be divided into two parts, an elastic lithosphere and a plastic lithosphere. The elastic lithosphere is defined as that part of the thermal lithosphere in which elastic stress can be maintained over geological times (Turcotte and Oxburgh, 1976). This is the case at temperatures of less than about 600°C. At temperatures between about 600°C and 1200°C the elastic stresses are relaxed by solid-state creep processes, but the net strain is still small so that the material does not flow. This is the plastic lithosphere.

Studies of the bending of the lithosphere under loads have derived thicknesses for the elastic lithosphere. Studies of the bending have been carried out for island chains (Walcott, 1970; Watts and Cochran, 1974), sea

mounts (Watts et al., 1975), and at ocean trenches (Hanks, 1971, Watts and Talwani, 1974; Caldwell et al., 1976). Typical thicknesses of the elastic lithosphere in the oceans are 25-40 km.

The characteristic length for the bending of the elastic lithosphere is the flexural parameter α. Typical values of the flexural parameter are 70-100 km. If the horizontal extent of a load is small compared to the flexural parameter, the bending rigidity of the lithosphere can support the load. Therefore, the load is not isostatically compensated. If the horizontal extent of the load is large compared with the flexural parameters, the load will be isostatically compensated. This problem has been considered in detail by Banks et al. (1977).

It has been suggested by Haxby et al. (1976) that the structure of the Michigan basin is the result of lithospheric flexure under localized loading. In Figure 4 the difference in depths between the Middle Ordovician Trenton Limestone and the Devonian Dundee formation in the southwest gradient of the basin are compared with an elastic flexure model. A cylindrical load with a radius of 95 km and an elastic lithosphere thickness of 32 km were assumed. Gravity studies confirm that the lithosphere is presently flexed under a central load. It has been argued by some authors that the progressive exposure of older strata away from the center of the basin indicates some stress relaxation since the basin formed about 400 million years ago. This stress relaxation has been treated using a viscoelastic model by Sleep and Snell (1976). An alternative explanation for the progressive exposure of older sediments is that several hundred meters of sediments have been eroded since the basin formed.

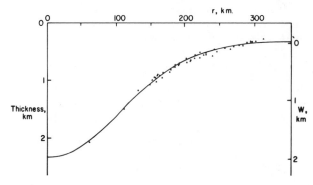

Fig. 4. The difference in depth between the Middle Ordovician Trenton Limestone and the Devonian Dundee Formation in the southwest quadrant of the Michigan basin as a function of distance from the center of the basin. The data points are from well logs, and the solid line represents the flexure of an elastic lithosphere with a thickness of 32 km under a cylindrical load with a radius of 95 km (after Haxby et al., 1976).

Faulting

Flexure may be prevented by faulting of the elastic lithosphere. Examples of where faulting has inhibited flexure are the passive margins of the Atlantic. The thermal subsidence of the oceanic lithosphere requires differential vertical movements of more than 3 km at the passive margins of a fully developed ocean. It has been shown by Turcotte et al. (1977) that the absence of significant gravity anomalies at these margins is strong evidence for vertical movement on the faults that develop during the initial opening of the ocean. A similar conclusion for the northern margin of the Mediterranean has been reached by Watts and Ryan (1976).

Some sedimentary basins may be fault-dominated early in their development but flexure-dominated during the later phases. As the rate of subsidence decreases, the marginal faults lock and lithospheric flexure can occur. The North Sea basin appears to be an example of this type of development.

In active tectonic regions the subsidence of very small sedimentary basins may be entirely fault controlled. Examples are the small pull-apart basins along the San Andreas fault in California. Various blocks within the Los Angeles basin appear to have subsided independently with vertical offsets occurring on various branches of the San Andreas system.

References

Ahrens, T. J., and G. Schubert, Gabbro-eclogite reaction rate and its geophysical significance, Rev. Geophys. Space Phys., 13, 383-400, 1975.

Artyushkov, E. V., Stresses in the lithosphere caused by crustal thickness inhomogeneities, J. Geophys. Res., 78, 7675-7708, 1973.

Banks, R. J., R. J. Parker, and S. P. Huestis, Isostatic compensation on a continental scale; local versus regional mechanisms, Geophys. J. R. astr. Soc., 51, 431-452, 1977.

Bott, M. H. P., and D. S. Dean, Stress at young continental margins, Nature, 235, 23-24, 1972.

Caldwell, J. G., W. F. Haxby, and D. L. Turcotte, On the applicability of a universal elastic trench profile, Earth Planet. Sci. Letters, 31, 239-246, 1976.

Collette, B. J., On the subsidence of the North Sea area, in D. T. Donovan, ed., Geology of Shelf Seas, Edinburgh, Oliver and Boyd, 15-30, 1968.

Falvey, D. A., The development of continental margins in plate tectonic theory, Australian Petrol. Explor. Asso. J., 14, 95-106, 1974.

Green, D. H., and A. E. Ringwood, A comparison of recent experimental data on the gabbro-garnet granulite-eclogite transition, J. Geology, 80, 277-288, 1972.

Hanks, T. C., The Kuril Trench-Hokkaido Rise system: large shallow earthquakes and simple models of deformation, Geophys. J. R. astr. Soc., 23, 173-189, 1971.

Haxby, W. F., D. L. Turcotte, and J. M. Bird, Thermal and mechanical evolution of the Michigan basin, Tectonophysics, 36, 57-75, 1976.

Joyner, W. B., Basalt-eclogite transition as a cause for subsidence and uplift, J. Geophys. Res., 72, 4977-4998, 1967.

Mareschal, J. C., and A. F. Gangi, Equilibrium position of a phase boundary under horizontally varying surface loads, Geophys. J. R. astr. Soc., 49, 757-772, 1977.

O'Connell, R. J., and G. J. Wasserburg, Dynamics of the motion of a phase change boundary to changes in pressure, Rev. Geophys., 5, 329-410, 1957.

Parsons, B., and J. C. Sclater, An analysis of the variation of ocean floor bathymetry and heat flow with age, J. Geophys. Res., 82, 803-827, 1977.

Sleep, N. H., Thermal effects of the formation of Atlantic continental margins by continental break-up, Geophys. J. R. astr. Soc., 24, 325-350, 1971.

Sleep, N. H., and N. S. Snell, Thermal contraction and flexure of mid-continent and Atlantic marginal basins, Geophys. J. R. astr. Soc., 45, 125-143, 1976.

Turcotte, D. L., J. L. Ahern, and J. M. Bird, The state of stress at continental margins, Tectonophysics, 42, 1-28, 1977.

Turcotte, D. L., and E. R. Oxburgh, Finite amplitude convection cells and continental drift, J. Fluid Mech., 28, 29-42, 1967.

Turcotte, D. L., and E. R. Oxburgh, Stress accumulation in the lithosphere, Tectonophysics, 35, 183-199, 1976.

Walcott, R. I., Flexure of the lithosphere at Hawaii, Tectonophysics, 9, 435-446, 1970.

Watts, A. B., and J. R. Cochran, Gravity anomalies and flexure of the lithosphere along the Hawaiian-Emperor seamount chain, Geophys. J. R. astr. Soc., 38, 119-141, 1974.

Watts, A. B., J. R. Cochran, and G. Selzer, Gravity anomalies and flexure of the lithosphere: a three-dimensional study of the Great Meteor seamount, northeast Atlantic, J. Geophys. Res., 80, 1391-1398, 1975.

Watts, A. B., and W. B. F. Ryan, Flexure of the lithosphere and continental margin basins, Tectonophysics, 36, 25-44, 1976.

Watts, A. B., and M. Talwani, Gravity anomalies seaward of deep sea trenches and their tectonic implications, Geophys. J. R. astr. Soc., 36, 57-90, 1974.

MECHANISMS OF SUBSIDENCE AT PASSIVE CONTINENTAL MARGINS

M. H. P. Bott

Department of Geological Sciences, University of Durham,
South Road, Durham DH1 3LE, England

Abstract. Some mechanisms for earlier graben
type subsidence and later regional downsagging
of shelf and slope at passive continental margins are reviewed. The gravity loading mechanism can explain thick sediment piles beneath
some deltas but fails to account for most of the
observed subsidence at margins except as a
contributory factor to other primary mechanisms.
Graben subsidence prior to continental splitting
or accompanying it can best be explained by
normal faulting affecting the upper part of the
crust in response to an applied tension of
uncertain origin. Later downsagging of the
margin may result from thermal contraction as
the underlying lithosphere cools back to normal
temperature following the thermal effects of the
split; this mechanism can produce subsidence
below sea level if the crust is thinned or its
density is increased prior to cooling. Geophysical observations indicate that thinning of
the adjacent continental crust does occur around
the time of continental splitting but by internal processes rather than by uplift and erosion.
One possibility is that the crust is thinned by
"necking" prior to continental splitting, and
this may be a viable mechanism in the narrow
graben setting provided that igneous intrusion
can adequately extend the brittle upper part of
the crust. Alternatively, the crust may be
thinned by seaward creep of continental crustal
material occurring rapidly directly after the
split, or subsequently at a slow pace.

Introduction

Passive continental margins undergo a history
of tectonic development of predominantly vertical type despite their location within plates.
The most obvious aspect is the strong subsidence
which affects particularly the shelf and slope
at rifted margins. Less obviously, there
appears to be a progressive widening of the
crustal transition with maintenance of approximate isostatic equilibrium. This paper reviews
some of the salient mechanisms which have been
recently suggested to explain subsidence at
rifted margins.

Four main stages can be recognized in the
tectonic development of a typical passive margin
of rifted type. (1) The rift valley stage,
which may not be ubiquitous, involves early
graben formation prior to or during continental
splitting and is possibly exemplified by the
present East African rift system. This stage
may be associated with domal uplift caused by
hot underlying upper mantle material, but such
doming may possibly be mainly restricted to hot
spot regions. (2) The youthful stage, lasting
about 50 Ma after the onset of spreading while
thermal effects of the split are dominant, is
exemplified by the Red Sea margins. This stage
is characterized by rapid regional subsidence of
the outer shelf and slope, but some graben subsidence may locally persist. (3) The mature
stage, during which more subdued regional subsidence may continue after the initial thermal
event ceases to be an important influence,
represents the present stage of development of
most of the Atlantic and Indian Ocean margins.
(4) The fracture stage, when subduction starts,
terminates the history of a passive margin.

Although passive margins vary greatly in
their style of tectonic development, two con-
trasting types of subsidence can be recognized
(Fig. 1): (1) Graben-type subsidence occurs
during the rift valley stage and may possibly
persist later. It is attributable to normal
faulting in the basement in response to crustal
stretching, possibly in association with doming;
(2) Subsidence by regional downwarping of the
margin without obvious fault control is characteristic of the youthful and mature stages, with
most rapid subsidence characteristically occurring at the start of spreading. Theories to
explain the mechanism of subsidence fall into
three main groups depending whether gravity
loading, temperature or stress is the primary
cause. There is still controversy concerning
the primary mechanism of regional downwarping,
with all three types of mechanisms probably con-
tributing. A stress mechanism possibly associated with a thermal event can best explain the
graben formation.

One approach to distinguishing between the

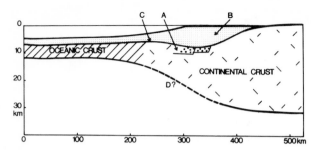

Fig. 1. Section across a passive margin showing some of the characteristic features of sediment and crustal structure: A = pre-split graben sediments, B = post-split sediments associated with flexural subsidence, C = the problematical position of the continent-ocean crustal contact beneath the sediments, D = the apparent gradational contact between deep continental and oceanic crust (after Bott, 1979).

hypotheses is the determination of the detailed history of subsidence from the stratigraphy and structure of the sediment pile. Complementary to this, the continent-ocean crustal contact needs to be located, and the nature of the crustal and upper mantle transition needs to be determined. Evidence of great importance in these respects has recently come from Leg 48 DSDP drilling and complementary geophysical investigations of the north Biscay margin as shown in Figure 2 (De Charpal et al., 1978). This "starved" margin has unusually thin post-split sediments so that the history of margin development can be studied without difficulty from pre-split times to the present. The continent-ocean crustal contact is located near the foot of the slope and has apparently remained undisturbed by faulting since formation. The age of the continental splitting here is Aptian (Lower Cretaceous). The following important inferences emerge from this study:

1. The pre-split sediments of Lower Cretaceous and earlier age now underlying the slope were deposited in epicontinental seas and overlie continental basement rocks; these sediments were presumably underlain by continental crust of normal thickness at the time of their formation.

2. Extensive normal faulting affected the pre-split sediments and the underlying basement at about the time of the split, producing a series of rotated and tilted fault blocks beneath the slope. Deep seismic reflections show that the faults are of listric type, flattening out into a plane of discontinuity at a depth of about 10 km below sea level. This faulting produced an extension of the upper crust beneath the slope amounting to about 10 percent.

3. The post-split sediments deposited on top of the fault blocks were apparently formed at about 2000 m depth, indicating that the margin here

subsided by about 2 km at the time of splitting rather than suffering uplift and erosion.

4. Seismic reflection lines show that the continental crust has been drastically thinned beneath the slope, the Moho being at about 13 km depth beneath the outer slope. It is reasonable to infer on isostatic grounds that the crustal thinning mainly occurred contemporaneously with the rifting stage.

5. Subsequently the slope has suffered further flexural subsidence amounting to over 2000 m at the outer slope.

This new evidence confirms the existence of the stages of rifting and flexural subsidence but adds the new factor of rapid subsidence and crustal thinning at the time of splitting. It is not yet clear whether this is of general applicability, as the relevant evidence at most passive margins is too deeply buried beneath thick sediment wedges.

Gravity Loading Hypotheses

The gravity loading hypothesis (Dietz, 1963; Walcott, 1972) attributes subsidence to sediment load. In its simplest form, it is based on local Airy isostasy. Suppose that the initial water depth is d and that densities of water, sediment and upper mantle are ρ_w, ρ_s, and ρ_m respectively. If the sea is filled by sediments, then the total thickness of marine sediments which may form with local isostatic adjustment is given by

$$t = d \cdot (\rho_m - \rho_w)/(\rho_m - \rho_s) \quad .$$

Putting $\rho_w = 1030$ kg/m^3 and $\rho_m = 3300$ kg/m^3, we see that a sediment thickness of about twice the initial depth can develop for sediment with mean density of 2150 kg/m^3, and of nearly three times the initial depth for mean density of

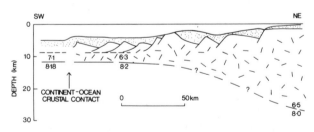

Fig. 2. Schematic crustal section across the north Biscay margin showing thin post-split sediments (stippled), listric normal faulting affecting basement and sediments of earlier age, the continent-ocean crustal contact near the foot of the slope, and thinned continental crust beneath the slope as revealed by seismic refraction results (velocities shown in km/s) (after De Charpal et al., 1978).

2550 kg/m^3. This shows that thick shelf successions with initial water depth of less than 200 m cannot form by this mechanism, but that substantial subsidence of the slope and rise may occur where thick piles of sediments are deposited such as at deltas. Figure 3b shows that a total thickness of sediment of about 14 km can form near the base of the initial slope.

A more sophisticated approach is to treat the lithosphere as a thin elastic plate and investigate its flexure in response to the sediment loading by elastic beam theory. Walcott (1972) used this technique to show the growth of a sedimentary lens at a continental margin, with particular relevance to deltas such as that of the Niger. Figure 3c shows a modification of Walcott's result using a sediment density of 2450 kg/m^3 (Walcott himself used the rather high density of 2800 kg/m^3). Over most of the original slope and adjacent oceanic crust, the resulting pile is closely similar to that predicted for classical isostasy (Fig. 3b), but the main distinction is that the downwarping extends about 150 km beyond the local sediment load. Thus, some significant subsidence of the shelf can be produced.

The gravity loading hypothesis thus gives a

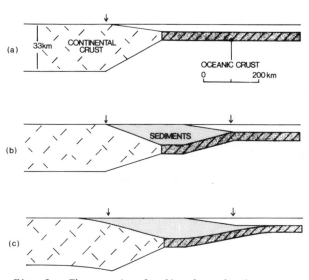

Fig. 3. The gravity loading hypothesis:
(a) The initial situation prior to loading, assuming a pre-existing 200 km wide transition between oceanic and continental crust following Walcott's (1972) model;
(b) The result of local Airy sediment loading, assuming density of the sediments of 2450 kg/m^3 and of the upper mantle of 3300 kg/m^3;
(c) The result of flexural loading, assuming that the lithosphere has a flexural rigidity of 2 x 10^{22} N m and that densitites are as in (b) (adapted from Walcott, 1972, with change of sediment density - after Bott, 1979).

viable explanation of thick delta wedges, with progressive seaward migration of slope and shelf over the continent-ocean crustal contact. However, Watts and Ryan (1976) show that neither local nor flexural loading can explain the substantial thicknesses of shallow water sediments typically observed on the shelves of passive margins. A "driving force" other than sediment loading is needed. Here the importance of the sediment loading effect is to increase the amount of subsidence caused by other mechanisms by a factor of between about two and three, depending on the mean sediment density. Thus, sediment loading appears to be a contributory factor in most subsidence, although it appears to be the primary cause only where great sediment volumes are deposited in initially deep water.

The Rifting Stage - Graben Formation

Modern concepts of graben formation stem from the hypothesis that a downward narrowing wedge of continental crust about 65 km wide can form by normal faulting in response to crustal tension and that this wedge subsides isostatically to form a rift valley between flanking uplifts (Vening Meinesz, 1950). This hypothesis fails in practice because the predicted crustal root is not observed and because graben tend to be narrower than 65 km. A viable mechanism of graben formation has recently been suggested in which the Vening Meinesz wedge subsidence concept is applied to the brittle upper 10-20 km of the crust as a whole (Bott, 1971; Fuchs, 1974; Bott, 1976).

This mechanism is based on the accumulating evidence that the continental crust and the lithosphere can be subdivided into a relatively strong upper elastic layer about 10-20 km thick which yields by brittle fracture, overlying a weaker lower layer which deforms by ductile flow (Artemjev and Artyushkov, 1971; Bott, 1971; Fuchs, 1974; Vetter and Meissner, 1979). This concept is supported by experimental and theoretical work starting with Griggs et al. (1960) and by lithospheric flexural studies (Walcott, 1970). The boundary between brittle and ductile layers is likely to be gradational and its depth will depend on the local geothermal gradient. It should be emphasized that the ductile layer is likely to be much stiffer than the asthenosphere and it may possess finite strength.

An outline of the current graben formation hypothesis is as follows (Fig. 4). A horizontal deviatoric tension is applied to the lithosphere and the tensile stress is enhanced in the brittle layer as a result of viscoelastic response in the ductile layer beneath (Kusznir and Bott, 1977). The brittle layer yields by normal faulting and a second converging normal fault may then develop where curvature is greatest or along a pre-existing line of weakness. This produces a downward narrowing wedge of

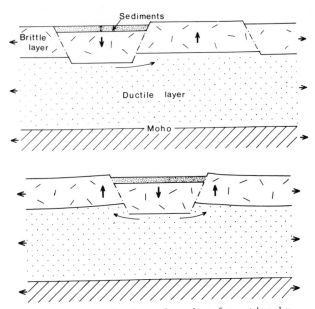

Fig. 4. The mechanism of graben formation by wedge subsidence affecting the upper continental crust with outflow in the lower crust:
(a) Subsidence compensated by horst uplift;
(b) Subsidence compensated by elastic upbending (after Bott, 1976).

brittle upper crust which can subside provided that the incremental loss of gravitational plus elastic strain energy is sufficient to overcome friction and other dissipation of energy. Bott (1976) showed that this mechanism can cause substantial subsidence provided that water pressure reduces friction on the faults. As the wedge subsides, outflow of material in the underlying ductile part of the crust causes uplift of the adjacent regions either by horst formation (Fig. 4a) or by elastic upbending (Fig. 4b). Complementary overall thinning of the ductile part of the crust must accompany subsidence so that the Moho may rise slightly beneath the graben vicinity, although this would not be by more than one or two kilometers in general. Within the setting of major rift systems, the above process may be complicated by associated igneous activity.

Bott (1976) suggested that the subsidence occurs by repetition of the following two stages: (1) Faulting stage, when the wedge subsides incrementally by normal faulting, accompanied by incremental drop in the tensile stress in the brittle layer with elastic shortening of the layer equal to the extension caused by faulting; (2) Stretching stage, when as a result of the faulting the tension in the ductile layer increases, causing progressive stretching and thinning as the stress is relieved in the ductile layer and is reapplied to the elastic layer above. This suggests that the graben subsidence should occur by repeated jerks

at a time interval depending on the time constant of viscoelastic response.

Graben of between 25 and 50 km width are predicted for realistic crustal parameters, with wider or narrower troughs being possible if there are basement weaknesses. A subsidence of about 5 km can occur for an initially 20 km wide trough under a maintained deviatoric tension of 50 MPa provided that sediments load the trough to its initial depth. In general, greater subsidence can occur for narrower troughs and for larger applied crustal tension.

This hypothesis can readily be applied to early graben formation along passive continental margins, either prior to continental splitting or continuing after it. If domal plateau uplift occurs as in East Africa, then the surface loading of the excess topography and the complementary upthrust of the underlying isostatic compensation cause horizontal deviatoric tension of about 20 MPa for a 2 km uplift. Kusznir and Bott (1977) showed that stress amplification as a result of viscoelastic flow beneath about 20 km depth will cause the deviatoric tension in the elastic layer to increase substantially. A domal structure thus provides one way in which the stress system favorable to graben formation can originate. Another possibility relevant to the northern Atlantic in the British region is that continental splitting farther south may give rise to the tensile stress system responsible for widespread Mesozoic graben formation.

Thermal Hypotheses of Post-Split Subsidence

The basic thermal hypothesis of Hsü (1965) has been theoretically developed in an elegant way by Sleep (1971, 1973). This hypothesis (Fig. 5) assumes that the continental lithosphere near the embryo margin is heated at the time of continental splitting. This causes reduction of the density of the lithosphere by thermal expansion and phase transitions with consequent isostatic uplift. After the initial split, as the ocean widens the lithosphere will cool and recover towards its initial elevation with a time constant of about 50 Ma. If, however, the crust has been thinned by surficial erosion at the time of uplift or by some other process, then the recovery on cooling will involve isostatic subsidence of the shelf which may be amplified by sediment load. This hypothesis would be expected to give rise to a smooth exponential decay type of subsidence reflected in sediment thicknesses, with time constant of about 50 Ma, but Sleep (1976) has shown that apparently jerky subsidence can be produced by superimposed eustatic sea level changes.

There are some factors which suggest that Sleep's hypothesis based on surficial erosion alone cannot be the sole explanation of regional subsidence at passive margins. The amount of crustal thinning, h, caused by erosion depends on the initial uplift and the time constants of

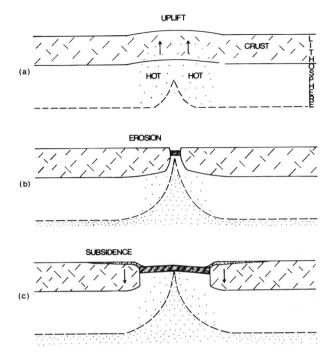

Fig. 5. The basic thermal hypothesis of Sleep (1971):
(a) Uplift following heating of the lithosphere;
(b) Initiation of new ocean accompanied by erosion of continental uplifted region, causing crustal thinning;
(c) Subsidence of continental margins as underlying lithosphere cools
(after Bott, 1979).
Note the nearly vertical edges of continental crust predicted by this model and that of Fig. 6.

cooling and erosion. Taking a maximum likely initial uplift of 2 km and an erosional time constant of 50 Ma, then h is about 4 km. The maximum possible sediment thickness to sea level which this will allow is given by

$$h.(\rho_m-\rho_c)/(\rho_m-\rho_s)$$

which is always less than h and typically about 1/2 h. Thus sediment thicknesses of up to about 2 km on the outer shelf can be explained, but unacceptably large amounts of supracrustal erosion would be needed to account for sediment thicknesses of 5 to 10 km which are commonly observed. Furthermore, a gap of about 50 Ma, during which erosion takes place, should intervene between the onset of spreading and the deposition of the first marine sediments. The evidence from the north Biscay margin of De Charpal et al. (1978) clearly shows that there has been no significant erosion and that no such time gap occurs. Lack of adequate erosion can be demonstrated at several other passive margins where sediments predating the split are preserved above the continental basement.

A modification of the simple thermal hypothesis is that the thermal event causes increase in the density of the lower continental crust by metamorphism (Falvey, 1974; Haxby et al., 1976), or by igneous intrusion (Beloussov, 1960; Sheridan, 1969), possibly followed by the basalt-eclogite transition (Ringwood and Green, 1966). According to Falvey (Fig. 6), rise in lithospheric temperature at the time of splitting causes greenschist facies rocks in the lower continental crust to be metamorphosed to amphibolite facies, with an increase in density of between 150 and 200 kg/m^3. This causes a slight thinning of the crust which to some extent counters the thermal uplift. As the continental lithosphere cools after spreading starts, subsidence below the initial level occurs because of the denser and slightly thinner crust. Taking the crustal thinning to be h, then the maximum sediment thickness to sea level which can result is given by

$$h.\rho_m/(\rho_m-\rho_s) \qquad .$$

Let us take an extreme example of an increase in density of 200 kg/m^3 affecting 15 km of lower crust, giving h of about 1 km. The maximum sediment thickness is then between about 3 and 4 km. Hence this version of the thermal hypothesis also fails to account for the great sediment thicknesses which occur on some margins. A similar difficulty faces the versions dependent on metamorphism or intrusion.

A version of the thermal hypothesis which avoids these difficulties is based on the thinning of the continental crust adjacent to the crustal contact by stretching of the crust

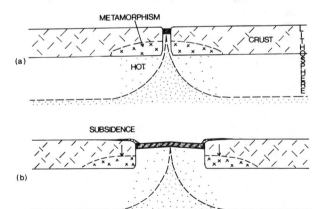

Fig. 6. The thermal hypothesis of Falvey (1974), omitting the rift stage:
(a) Heating of the continental lithosphere prior to and during split causes metamorphic transition of lower crust from greenschist to amphibolite facies, raising the mean crustal density;
(b) Subsidence of continental margins as underlying lithosphere cools
(after Bott, 1979).

just before the split or by outflow of continental crustal material just after the split (Fig. 7). Possible mechanisms for such crustal thinning are discussed in the following section of the paper. The evidence from the north Biscay margin indicates that such crustal thinning did occur at about the time of the split, on a great enough scale to counteract any thermal uplift and furthermore to cause an immediate subsidence of about 2000 m. Subsequently, the adjacent thinned continental and oceanic crust at this margin underwent a further subsidence of over 2000 m as the underlying lithosphere cooled back towards its normal thermal state. Thus, there was here an overall subsidence of over 4 km without any significant sediment loading, about half of this being the immediate response to crustal thinning and half of it due to later cooling of the lithosphere. If this same margin had been loaded to sea level by sediments, the overall subsidence would have been about 10 km, a value typical of several passive margins.

The thermal hypotheses thus provide an elegant explanation of exponentially decaying post-split subsidence at passive margins provided that crustal thinning occurs at the time of the split or shortly afterwards by processes other than erosion. An additional smaller contribution to the subsidence may possibly arise through metamorphism or igneous intrusion into the underlying continental crust. The remaining problem is to understand how the crustal thinning has taken place.

Crustal Thinning Hypotheses

Recent observations show that the continent-ocean contact at sediment-starved rifted margins typically lies at the foot of the slope or even oceanward of it. The continental crust beneath the slope thins towards the crustal contact where it is about equal in thickness to the adjacent oceanic crust. Examples of such a situation include the north Biscay margin (De Charpal, 1978), the margins on both sides of Rockall Trough and to the west of the Rockall microcontinent (Roberts, 1975; Bott, 1978) and the southeastern Greenland margin (Featherstone et al., 1977). The north Biscay study shows that the main crustal thinning probably occurred at about the time of the continental split. Gravity studies show that a similar gradational crustal transition occurs generally beneath more typical passive margins where post-split sediments are thick, although the location of the continent-ocean crustal contact is usually uncertain. We assume here, as a working hypothesis, that the mechanism of crustal thinning applicable to starved margins also affects the sediment swamped margins, the only major tectonic distinction being the degree of sediment loading. The three main mechanisms which have been suggested to explain such crustal thinning

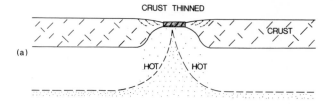
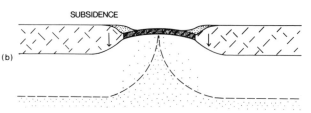

Fig. 7. The thermal hypothesis with crustal thinning by processes other than uplift and erosion:
(a) Crustal thinning and rifting at the time of continental splitting;
(b) Subsidence of continental margins as underlying lithosphere cools.

are: (1) phase transition of gabbro to eclogite causing the Moho to migrate upwards, (2) stretching of the crust and (3) continental crustal creep towards the margin after the split. The gabbro-eclogite hypothesis (Collette, 1968) has not received much recent support because of the unlikelihood that the Moho is a phase boundary, but it should still be borne in mind as a possibility. The other two mechanisms are described below.

Thinning of the Crust by Stretching

It has been suggested that extreme thinning of the continental crust within a rift valley setting can occur by plastic necking (Artemjev and Artyushkov, 1971; Kinsman, 1975). If such a rift zone subsequently splits to form complementary passive margins, then the region of thinned crust would give rise to a zone of crustal transition beneath the slope. More recently, McKenzie (1978) has suggested, with reference to North Sea subsidence rather than to passive margins, that basin subsidence is initiated by stretching of the continental crust, causing a rapid initial subsidence at the time of stretching followed by slower subsidence in response to cooling from the raised temperatures produced by the stretching.

Can such pre-split stretching of continental crust and lithosphere account for the observed crustal thinning at passive margins such as the north Biscay margin? The upper part of the crust is brittle and the lower part is probably ductile. If stretching occurs before continental splitting, then the same overall extension must affect both parts of the crust. Thus the fractional thinning of the crust averaged over the affected zone must equal the average fractional extension, assuming plane strain. This applied whether or not the faults are listric. The problem of the stretching hypothesis is to explain how the brittle upper part of the crust can be stretched by a factor of two or more. This might come about either by an extreme criss-cross pattern of normal faulting or by bodily disruption of the brittle layer with voids between the pulled-apart blocks possibly being invaded by magma. We should be able to find evidence for such faulting or bodily disruption where it has occurred.

The north Biscay margin is probably the only one for which adequate evidence is available to carry out such an analysis. Here there is no indication of bodily disruption or of extensive igneous invasion of the upper crust. De Charpal et al. (1978) use the normal faulting and associated block tilting to show that the extension of the upper crust between shelf edge and continent-ocean contact is about 10 percent. On the other hand, the average thinning of the lower ductile part of the crust beneath the slope is more than 50 percent. Thus, the stretching mechanism does not seem adequate to account for more than about a quarter of the observed crustal thinning. One possible explanation is that the thinning and extension of the ductile lower crust is restricted to the present slope region but that the stretching of the brittle upper crust extends much farther back into the continental region, the two layers being decoupled from each other well back into the continent. Alternatively, crustal material must have been removed from the region beneath the slope by some other mechanism.

If the north Biscay pattern of stretching can be regarded as typical of passive margins, the evidence suggests that stretching may not even be the main process of crustal thinning, although it must contribute.

Crustal Thinning by Creep at Margins

A stress-based hypothesis which may account for subsidence at passive margins appeals to thinning of the continental crust near the margin by progressive creep of ductile middle and lower crustal material towards the sub-oceanic upper mantle (Bott, 1971, 1973). The flow is driven by the release of gravitational energy as the continent-ocean crustal transition becomes progressively more gradational. The hypothesis depends on the ability of the lower and middle continental crust to flow significantly by steady state creep while the overlying elastic layer subsides by elastic flexure or by normal faulting.

Bott and Dean (1972) showed that passive margins are associated with a differential stress system as a result of unequal topographic loading across the margin and associated upthrust of the low density continental crust in isostatic equilibrium with the oceanic region. The additional loading and compensatory upthrust on the continental side effectively squeezes the continental crust, causing a horizontal deviatoric tension in the continental crust of up to 20 MPa, or a stress difference of 40 MPa. The deviatoric tension peaks in the middle of the crust, decreasing to zero at the surface and at the Moho. Kusznir and Bott (1977) have recently shown that if the lower crust is treated as viscoelastic, then stress differences in the upper elastic layer will be increased. The main point is that a stress system does exist in the continental crust adjacent to a passive margin due to the unequal loading, and that this will tend to drive lower continental crustal material towards the sub-oceanic upper mantle.

According to this hypothesis (Fig. 8), the crustal thinning will cause isostatic subsidence of the shelf and possibly slope, which will be accentuated by the sediment load. At the same time there will be a progressive broadening of the transition between oceanic and continental crust, in contrast to the predictions of the thermal hypotheses.

The rate of subsidence is likely to be controlled by both temperature and superimposed stresses. Solid state creep is a thermally activated process, so that the most rapid subsidence would be expected to occur at and shortly after the initiation of spreading when lithospheric temperatures are high. Thus an approximately exponential decay of subsidence with time might be expected for this hypothesis, making distinction from the thermal hypotheses difficult on these grounds. The other controlling factor is likely to be the external stress system associated with the plate driving mechanisms or other sources such as membrane stresses (Turcotte, 1974). This will be superimposed on the nearly steady stress system associated with the unequal loading effect. Rapid subsidence would be encouraged by a superimposed tension but inhibited by a superimposed compression. Thus, subsidence according to this mechanism is likely to be partly thermally and partly stress controlled.

The crustal creep mechanism may contribute to passive margin subsidence at two stages. Firstly, rapid outflow and collapse of the newly split continental crust into the hot upwelling mantle may cause crustal thinning and subsidence just after the time of initial splitting; it is possible that such a mechanism may contribute to the crustal thinning at the north Biscay margin during, say, the first 5 Ma after the split. Secondly, at a later stage, the mechanism may contribute to the subsidence normally attributed to lithospheric cooling. In particular, local regions of differential subsidence such as the Hatton Basin on the Rockall microcontinent (Matthews and Smith, 1971; Bott, 1972) can be accounted for best by the creep mechanism. The main problem associated with the hypothesis

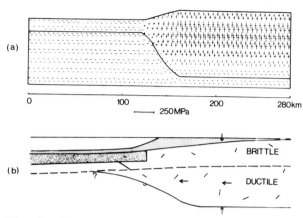

Fig. 8. The crustal flow hypothesis of Bott (1971):
(a) Magnitude and direction of the principal compressive stresses arising from differential gravitational body forces across a passive margin, the intermediate principal stress being everywhere perpendicular to the model (after Bott and Dean, 1972).
(b) Progressive gradation of the crustal transition caused by seaward flow of the lower and middle continental crust in response to the stress system. This causes thinning of the continental crust with consequent isostatic subsidence, and complementary uplift of the rise (after Bott, 1979).

is in locating the outflowed continental material within the adjacent oceanic lithosphere. During the earlier stage, it may possibly become incorporated in the magma injected to form new oceanic crust and part of it may remain located in the sub-oceanic upper mantle beneath the rise. There is no direct evidence for the presence of such material. However, the low density continental material would be expected to cause some complementary uplift of the continental rise, and Rona (1974) presents evidence which might be interpreted in this way.

Conclusions

It has been shown that three main types of mechanisms probably contribute to subsidence at passive margins. The sediment loading mechanism may account for thick sediment piles at some deltas but is apparently not the primary cause of most shelf subsidence; however, sediment loading is a contributory factor in subsidence originating by other processes, increasing the possible sediment thickness by a factor of between two and three. The thermal mechanism gives an elegant explanation of the apparent exponential decay of the rate of subsidence with time after the initial thermal event (at continental splitting) but requires accompanying crustal thinning by processes other than erosion or metamorphism if shelf sediment piles of more

than about 4 km are to be accounted for. The stress-based mechanisms may explain the thinning of the continental crust adjacent to the margin either by stretching at the time of continental splitting or by oceanward creep of continental crustal material.

The north Biscay margin provides a model of the subsidence where the influence of sediment loading is minimal. Two main stages of subsidence have occurred. The first stage, possibly straddling the time of continental splitting, involves a 10 percent stretching of the upper continental crust beneath the slope by listric normal faulting, an extreme thinning of the continental crust so that it is about equal in thickness to the oceanic crust at their contact near the foot of the slope, and a rapid subsidence of about 2000 m, at the crustal contact. This stage can be attributed to stress-based mechanisms of crustal thinning, such as by stretching of the crust and possibly by outflow of ductile lower continental crust into the newly formed sub-oceanic region. The second stage has involved the post-split flexural subsidence of the region beneath the slope with the continental and oceanic crusts at their contact subsiding together by about 2400 m at the same rate. This can best be mainly attributed to thermal subsidence on cooling of the oceanic region and adjacent continental region beneath the slope after the thermal effects of the split, but some oceanward creep of continental material may contribute.

It is suggested, as a working hypothesis, that the only tectonic distinction between the north Biscay margin and a typical passive margin is the additional subsidence caused by sediment loading at the latter. If adequate sediment supplies had been available to load the north Biscay margin to sea level, the overall subsidence at the continent-ocean contact would be between about 9 and 12 km with a comparable maximum thickness of the wedge of post-split sediments.

Acknowledgments

This paper is partly based on an earlier review (Bott, 1979) published by the American Association of Petroleum Geologists, and in particular Figures 1, 3-6, and 8 are reproduced from the paper. The author is grateful to the AAPG for permission to re-use this material.

References

Artemjev, M. E., and E. V. Artyushkov, Struc- and isostasy of the Baikal rift and the mechanism of rifting, J. Geophys. Res., 76, 1197-1211, 1971.
Beloussov, V. V., Development of the earth and tectogenesis, J. Geophys. Res., 65, 4127-4146, 1960.
Bott, M. H. P., Evolution of young continental

margins and formation of shelf basins, Tectonophysics, 11, 319-327, 1971.

Bott, M. H. P., Subsidence of Rockall Plateau and of the continental shelf, Geophys. J.R. astr. Soc., 27, 235-236, 1972.

Bott, M. H. P., Shelf subsidence in relation to the evolution of young continental margins, in D. H. Tarling and S. K. Runcorn, eds., Implication of continental drift to the earth sciences, 2, Academic Press, London and New York, 675-683, 1973.

Bott, M. H. P., Formation of sedimentary basins of graben type by extension of the continental crust, Tectonophysics, 36, 77-86, 1976.

Bott, M. H. P., The origin and development of the continental margins between the British Isles and southeastern Greenland, in D. R. Bowes and B. E. Leake, eds., Crustal evolution in northwestern Britain and adjacent regions, Geol. J. Spec. Issue 10, Seel House Press, Liverpool, 377-392, 1978.

Bott, M. H. P., Subsidence mechanisms at passive continental margins, in Geological and geophysical investigations of continental margins, Am. Assoc. Petrol. Geol., Mem. 29, 3-9, 1979.

Bott, M. H. P., and D. S. Dean, Stress systems at young continental margins, Nature (Phys. Sci.), Lond., 235, 23-25, 1972.

Collette, B. J., On the subsidence of the North Sea area, in D. T. Donovan, ed., Geology of shelf seas, Oliver and Boyd, Edinburgh and London, 15-30, 1968.

De Charpal, O., P. Guennoc, L. Montadert, and D. G. Roberts, Rifting, crustal attenuation and subsidence in the Bay of Biscay, Nature, Lond., 275, 706-711, 1978.

Dietz, R. S., Collapsing continental rises: an actualistic concept of geosynclines and mountain building, J. Geology, 71, 314-333, 1963.

Falvey, D. A., The development of continental margins in plate tectonic theory, J. Australian Petrol. Explor. Assoc., 14, 95-106, 1974.

Featherstone, P. S., M. H. P. Bott, and J. H. Peacock, Structure of the continental margin of southeastern Greenland, Geophys. J.R. astr. Soc., 48, 15-27, 1977.

Fuchs, K., Geophysical contributions to taphrogenesis, in J. H. Illies and K. Fuchs, eds., Approaches to taphrogenesis, Schweitzerbart, Stuttgart, 420-432, 1974.

Griggs, D. T., F. J. Turner, and H. C. Heard, Deformation of rocks at 500° to 800°C, Geol. Soc. Am. Mem. 79, 39-104, 1960.

Haxby, W. F., D. L. Turcotte, and J. M. Bird, Thermal and mechanical evolution of the Michigan Basin, Tectonophysics, 36, 57-75, 1976.

Hsü, K. J., Isostasy, crustal thinning, mantle changes, and the disappearance of ancient land masses, Am. J. Sci., 263, 97-109, 1965.

Kinsman, D. J. J., Rift valley basins and sedimentary history of trailing continental margins, in A. G. Fischer and S. Judson, eds., Petroleum and global tectonics, Princeton Univ. Press, 83-126, 1975.

Kusznir, N. J., and M. H. P. Bott, Stress concentration in the upper lithosphere caused by underlying visco-elastic creep, Tectonophysics, 43, 247-256, 1977.

Matthews, D. H., and S. G. Smith, The sinking of Rockall Plateau, Geophys. J.R. astr. Soc., 23, 491-498, 1971.

McKenzie, D., Some remarks on the development of sedimentary basins, Earth and Planet. Sci. Lett. (Neth.), 40, 25-32, 1978.

Ringwood, A. E., and D. H. Green, An experimental investigation of the gabbro-eclogite transformation and some geophysical implications, Tectonophysics, 3, 383-421, 1966.

Roberts, D. G., Marine geology of the Rockall Plateau and Trough, Phil. Trans. R. Soc. Lond., Ser. A, 278, 447-509, 1975.

Rona, P. A., Subsidence of Atlantic continental margins, Tectonophysics, 22, 283-299, 1974.

Sheridan, R. E., Subsidence of continental margins, Tectonophysics, 7, 219-229, 1969.

Sleep, N. H., Sensitivity of heat flow and gravity to the mechanism of seafloor spreading, J. Geophys. Res., 74, 542-549, 1969.

Sleep, N. H., Thermal effects of the formation of Atlantic continental margins by continental breakup, Geophys. J.R. astr. Soc., 24, 325-350, 1971.

Sleep, N. H., Crustal thinning on Atlantic continental margins: evidence from older margins, in D. H. Tarling and S. K. Runcorn, eds., Implications of continental drift to the earth sciences, 2, Academic Press, London and New York, 685-692, 1973.

Sleep, N. H. Platform subsidence mechanisms and "eustatic" sea-level changes, Tectonophysics, 36, 45-56, 1976.

Turcotte, D. L., Membrane tectonics, Geophys. J.R. astr. Soc., 36, 33-42, 1974.

Vening Meinesz, F. A., Les grabens africains, résultat de compression ou de tension dans la croûte terrestre?, Bull. inst. R. Colon. Belge, 21, 539-552, 1950.

Vetter, U. R., and R. O. Meissner, Different creep properties of adjacent continental and oceanic lithospheres, Tectonophysics, 1979, in press.

Walcott, R. I., Flexural rigidity, thickness, and viscosity of the lithosphere, J. Geophys. Res., 75, 3941-3954, 1970.

Walcott, R. I., Gravity, flexure, and the growth of sedimentary basins at a continental edge, Bull. Geol. Soc. Am., 83, 1845-1848, 1972.

Watts, A. B., and W. B. F. Ryan, Flexure of the lithosphere and continental margin basins, Tectonophysics, 36, 25-44, 1976.

THE ORIGIN OF VERTICAL CRUSTAL MOVEMENTS WITHIN LITHOSPHERIC PLATES

E. V. Artyushkov

Institute of Physics of the Earth, B. Gruzinskaya 10, Moscow D-242, USSR

A. E. Shlesinger

Geological Institute of the USSR Academy of Sciences, Pyjewsky per. 7
Moscow, USSR

A. L. Yanshin

Institute of Geology and Geophysics of the Siberian Branch
of the USSR Academy of Sciences, Universit. prospect, 3, Novosibirsk 90, USSR

Abstract. At the core-mantle boundary the authors postulate a phase transformation from solid state into a liquid and denser state with the acquisition of metallic properties. This process leads to the formation of unmolten differentiates that are lighter than the lower mantle. Light lower mantle material is released with a periodicity of about 200 Ma and rises through channels (about 100 km in diameter) that heat the surrounding mantle and facilitate further passage of additional light lower mantle material. Such material then emerges in the upper mantle, differentiates further, and spreads below the lithosphere or even below the earth's crust. Thickness variations in the lithosphere lead to the formation of "traps" that preferentially catch the light material where the lithosphere is thin and the formation of "antitraps" where the lithosphere is thick. As such traps are located at the base of the moving plates of the earth, they would catch light material whenever they can get it from a deep channel rising from the lower mantle. The trapping of hot light material leads to the formation of broad uplift areas of minor relief (shields) when only minor amounts of light material can be intercepted repeatedly from relatively distant channels. However, in tectonically more active areas where larger traps are located near the ascent channels, the upper mantle is heated further and leads to broad uplift zones such as the western U.S. or the Baikal area. Low-lying shield areas may also be reactivated by vigorous new entrapment of light mantle material as in the Tien Shan Mountains.

Sedimentary basins are formed by any combination of three processes: (1) mantle consolidation in the area of a former trap, (2) conversion by phase transition of the lower part of

the basaltic layer of the crust into eclogite, or (3) transition of a part of the overlying basalt into garnet granulite.

Large masses of heated light material supplied to the base of the lithosphere beneath the sedimentary basins on the platforms eventually displace colder upper mantle material to come into contact with the basaltic layer of the crust, there leading to a quick transition from basalt into garnet granulite and into eclogite. Such a history is reflected in the history of larger inland seas (e.g., south Caspian Basin) where an earlier platform sequence is separated by an unconformity suggesting uplift and followed by renewed and intensified subsidence.

Introduction

The earth's crust is in perpetual movement. Horizontal movements are of the greatest amplitude. Sea-floor spreading and continental drift involve lithospheric plate displacements of thousands of kilometers. Such displacements are accompanied by intense deformation along plate boundaries. However, within plates, horizontal movements virtually do not cause lithospheric deformations.

Nevertheless and obviously, vertical movements within plates are manifest. These cause the relief of the earth's surface and thus are best studied on the continents. Therefore in this paper we are going to deal with continental areas only.

Substantial areas of the lithospheric plates are characterized by secular low intensity vertical movements. Such areas constitute the continental platforms. Within them are shields which are areas where positive movements were predominant and where denudation brought the crystalline basement to the surface. In contrast to shields, platform sedimentary basins are areas of predominantly negative movements in which great volumes of sediments accumulated.

In tectonically active areas, the rate of vertical movements of lithospheric plates is one order or several orders of magnitude higher. There we find rising areas, that is, areas of tectonic activization or of epiplatform orogenesis with mountainous relief and areas of intense subsidence that formed deep depressions of the type of inland seas.

A characteristic of vertical movements in many areas is their unidirectional - either upward or downward - movement which lasts for very long periods of time that sometimes exceed a billion years. Thus the Baltic shield, for instance, experienced uplift for over 1500 Ma.

The origin of vertical crustal movements within lithospheric plates is still not properly understood. They are usually associated with horizontal movements of the lithospheric plates. However, as we demonstrate further on, such an explanation is not likely to be valid.

Vertical movements may also be explained by flows in the mantle. Van Bemmelen (1933) apparently formulated for the first time an idea where vertical crustal movements have been associated with the emergence of the light material from the lower mantle into the upper mantle. In this process uplifts are being formed on the surface of the earth over ascending flows and subsiding areas under descending flows. However, because of the drastic drop in the viscosity in the asthenosphere, to values $\leq 10^{20}$ poise (Artyushkov, 1971), such a mechanism should have little effect. Under the influence of vertical movements in deeper layers, the asthenospheric matter should spread rapidly sideways. As a result, only comparatively small displacements $\leq 10^2$m can originate on the surface of the earth.

A more effective mechanism for vertical movements may be the ascent of the light material into the very top layers of the earth. Approaching the crust such material forms the roots of positive structures, which are isostatically uplifted (Beloussov, 1962, 1966). It has been assumed in these papers that the fusion of the light material takes place in the underlying asthenosphere. This mechanism has been suggested as an explanation of vertical movements in geosynclines. However, in addition to uplifts in geosynclines, ascending movements take place also in many other areas, e.g., on the shields of platforms. From the point of view now under discussion, it is necessary to assume that under all these areas, partial melting of the asthenosphere of varying intensity is taking place. And yet, judging from geothermal (Lubimova and Smirnov, 1977) and seismological data (Alexeev and Ryaboy, 1976), the asthenospheric layer has a high temperature only under tectonically active areas. Platforms, however, are far from the melting point.

A multiple recurrence of vertical movements in the same areas can be explained only if the lithospheric plates are not at all horizontally displaced and melting occurs repeatedly in the underlying mantle. In that case, first of all, the light material molten from the asthenosphere carries away a substantial amount of radioactive elements into the crust. Therefore, any subsequent new melting becomes rather improbable. Furthermore, after the separation of the light material, its volume in the asthenosphere diminishes, so that new segregation of light material may take place only in considerably smaller amounts.

The nature of descending vertical movements has been investigated even less. An obvious process may be thermoelastic contraction of the crust along with its cooling (Foucher, 1976). This mechanism, however, produces only a very slight sinking $\leq 1-2$ km.

Another mechanism - oceanization - has been suggested for the formation of deep basins like inland and marginal seas (or even of oceans) with a crust greatly reduced in thickness replacing the former continental crust (Beloussov, 1968). It is assumed that the crust becomes heavier than the underlying mantle and sinks into the mantle as the result of the intrusion into the crust of a great number of ultrabasic bodies. In fact, such a process is impossible (Lyustikh, 1963) because the average density of the crust is much lower than the density of the peridotite of the mantle. Therefore, a mixture of the bodies of the usual crustal and mantle composition always remains lighter than the mantle matter and consequently cannot sink in it.

Density Differentiation of the Earth's Primary Substance

The fundamental characteristic of the internal structure of the earth is the existence of a liquid core that is denser than the surrounding mantle. Right after its formation, the primary earth represented a mixture of substances of various densities (Schmidt, 1960; Levin, 1972). The liquid and denser core had to become segregated from the primary substance of the earth (or better from the primary mixture of substances) in the process of the evolution of the planet.

Today the density differentiation of the primary substance of the earth occurs at the boundary between the core and the mantle (Artyushkov, 1968, 1970). Thus the lower mantle may contain inclusions of core substance in a solid state; whereas under the lower mantle in the outer core this substance is in a liquid phase. The isotherm of melting coincides with the core-mantle boundary. It appears most probable that at the core-mantle boundary there occurs not only simple melting of the core substance but also a phase transformation from a solid state into a liquid and denser state accompanied by the acquisition of metallic properties. If the lower mantle consists of oxides and the core contains a large amount of iron, a metallization is most probably experienced by FeO (Dubrovsky and Pankov, 1972). The pressure at which such a transformation is taking place should decrease with a rise in temperature (Magnitzky, 1967). Consequently, along with the heating of the earth as a result of radioactive decay, the surface in the lower mantle on which this transformation occurs should also rise.

As a result of the melting of the core substance, the rock skeleton of the lower mantle is destroyed and the rock itself changes into a partly melted and effectively liquid state. The heavier substance of the mantle sinks in the melt and joins the liquid core. After this differentiation in the lower mantle, a mixture of unmelted substances remains that is lighter than the lower mantle which still contains the heavy substance of the core. As a result of a convective instability, this light material penetrates into the overlying mantle and finally gets into the top layers of the earth. It is most probable that the light material evacuated from the core-mantle boundary since the existence of the earth forms the top layer of the earth (~1000 km thick), i.e., the upper mantle in the first approximation. The layer below – the lower mantle – represents the still un-differentiated substance of the earth.

The ascent of the light material in the lower mantle must be discontinuous not only in space, but also in time. The rocks are characterized by a strongly nonlinear stress-strain dependence. The stresses are created in the lower mantle because of the non-uniform distribution of the light material underneath it. These stresses are intensified with an increase of its volume. As a result, the ascent of the light material in the lower mantle should begin only after its accumulation in a sufficiently large amount. This concept permits us to explain the existence of tectonic cycles such as a sharp activization of tectonic movements with a periodicity of about 200 Ma.

With an injection of the light material into the lower mantle due to a convective instability the size of its blocks will be larger and of the order of the thickness of the light material layer accumulated under the mantle during one tectonic cycle. Assuming that the average differentiation during the period of the existence of the earth proceeds uniformly, an addition to the upper mantle during a 200 Ma interval will be approximately one-twentieth of its mass and corresponds to the thickness of the light material layer formed under the lower mantle of $h \sim 100$ km.

The ascent of the light material in the lower mantle is accompanied by a high release in the potential energy. It is

$$U = \int_{r_1}^{r_2} \Delta\rho g dr \qquad (1)$$

per unit volume of the light material. In this equation $\Delta\rho$ is the difference between the density of the surrounding mantle and the light material; r_1 = the radius of the core, r_2 = the radius of the lower mantle top. With $r_2 - r_1 \sim 2000$ km, $\Delta\rho \sim 100-200$ kg/m^3, g=10 m/s^2.

$$U \sim 4.18 \text{ GJ/m}^3 \qquad (2)$$

This energy is spent on overcoming the forces of viscous friction at the emergence of the light material and, finally, becomes transformed into heat.

After the block of light material passes through the lower mantle – being as an average at rest – a vertical zone of higher temperature remains along its passage and consequently, also a zone of lower viscosity. It represents a "channel" which facilitates ascent into the upper mantle of the next blocks of light material. The emergence of the light material will tend toward such channels along which a greater amount of material had already passed and which, consequently, are of a lower viscosity. The diameter of these channels should be of the same size order as the blocks injected into the lower mantle, i.e., $\sim 10^2$ km.

Because of the low thermal diffusivity of the lower mantle, $\kappa \sim 10^{-6}$ m^2/s, the heat has no time to spread far beyond the walls of the channel. During one tectonic cycle, i.e., during the time $t \sim 200$ Ma, areas become heated up to a distance of ≤ 100 km from the walls. Consequently, all the gravity energy is released in a relatively small volume in the channel or near it. It is being spent on the heating of the lower mantle substance and of the light material. When a volume of light material greatly exceeding the volume of the channel is passing through the channel, the main share of the released energy is consumed by the light material. With a heat capacity $C_p \sim 4.18$ MJ/m^3 (Zharkov, 1962) and a gravitational energy release (2), the temperature rise of the light material after its passage through the lower mantle comes to

$$\Delta T \sim 1000° \qquad (3)$$

At such a temperature the density of the light material should be below the average density of the upper mantle. Therefore having passed the channel in the lower mantle, it continues to emerge in the upper mantle as well and penetrates into its top layers pushing away previously located colder substances of the

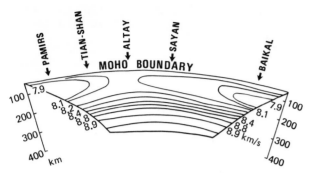

Fig. 1. Seismic section of the upper mantle along the Pamirs-Baikal profile (Alexeev et al., 1971).

asthenosphere. An accumulation of such material is being recorded at the present time under many tectonically active areas. Figure 1 shows a seismic section across Central Asia. As shown on this figure, at depths averaging from 50 to 250 km, substantial heterogeneities are observed in the asthenosphere. A sharp drop in the velocity of longitudinal waves takes place under the Pamirs and partly under Tien Shan, as well as under Baikal and Eastern Sayan. Under the Altai and Western Sayan the velocities are substantially higher, but they still remain lower than at the same depths under platforms, where they usually are 8.5-8.8 km/s. A strong drop in the velocities is probably associated with a higher temperature of the mantle approaching the melting point.

Anomalous heating of the asthenosphere is also indicated by geothermal data, according to which the temperature here at the Moho comes to 1000-1200°C (Smirnov, 1977). This is also indicated by an intense basaltic volcanism characteristic for many tectonically active areas (Holmes, 1969; Grachev, 1977). For comparison, it may be noted that the temperature at the same depths under stable platform areas is at least 400-500° lower. Let us see whether the described heterogeneities could be the result of thermal convection in the mantle. The velocity of thermal convective currents under the lithosphere in any case should not be below the rate of drift of the lithospheric plates themselves, i.e., ~0.1 m/a. At such a velocity a hot substance rising from depth would then spread under the lithosphere over large distances gradually departing from the area of its ascent and from the uplift area on the surface of the earth overlying it. At the same time, intensely heated areas in the asthenosphere are distinctly "tied-up" to tectonically active areas on the surface of the earth and are not extending far beyond their limits. Thus, for instance, the roof of the electrical high conductivity layer under the Baikal arch uplift is at a depth of 80 km and under the Baikal rift itself - at a depth of 40 km (Vanyan and Harin, 1967; Bulmasov

et al., 1968). As it extends to the Siberian platform adjacent to the rift from the northwest, this surface sharply sinks to a depth of 200 km. The Baikal arch uplift began its formation in the Oligocene (Florensov, 1960), i.e., not less than 30 Ma ago. If the heated material in the thermal convection cell would spread sideways at a rate of ~0.1 m/a, the inhomogeneity in the mantle would have extended beyond the limits of the uplift to a distance of 3000 km.

In this way, judging by the data on the structure of the asthenosphere, there is no place in the mantle for thermal convection in the form of large cells reaching the lithospheric layer at the top. It should be noted that in general, thermal convection becomes much more difficult in the presence of density differentiation in the earth.

In the ascent areas of a highly heated light material into the asthenosphere, it is possible to expect that its partial melting is sometimes accompanied by a density differentiation. (In the lower part of the upper mantle, the melting is much less probable, because the melting temperature of rocks rapidly increases with pressure.) The lightest components of the light material, segregated by differentiation, during the emergence, can reach the base of the lithosphere or else of the earth's crust. In tectonically active areas directly under the crust, there are actually thick lenses of the mantle with strongly reduced velocities of longitudinal waves $V_p \sim 7.2$-7.8 km/s. These are often referred to as "anomalous mantle" or a "low-velocity mantle" that differs from a normal mantle with velocities $V_p \sim 8.1$-8.3 km/s that is usually located under the crust of stable areas. The temperature of the anomalous mantle is $T \sim 1000$-$2000°C$. A well-known example is the western part of the United States of America, where under a crust ~30-40 km thick, there is a layer of an anomalous mantle ~30 km thick (Fig. 2). Large masses of an anomalous mantle are located under the mid-oceanic ridges (Talwani et al., 1965).

Spreading of an Anomalous Mantle along the Base of the Lithosphere

The viscosity of the low-viscosity layer in stable areas is $\eta \sim 10^{19}$-10^{20} poise (Artyushkov, 1971a). With a rise in the mantle temperature by each 100°, the viscosity drops approximately by one order of magnitude (Artyushkov, 1973). In tectonically active areas, the asthenosphere temperature is higher than in stable areas by 300-500°. Consequently, the viscosity of the asthenosphere there is

$$\eta \lesssim 10^{17} \text{ poise} \qquad (4)$$

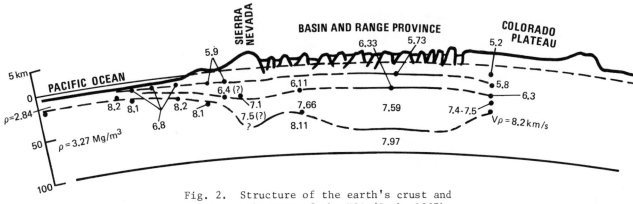

Fig. 2. Structure of the earth's crust and
mantle in the west of the USA (Cook, 1967).

The viscosity of the continental lithosphere is
very high:

$$\eta \lesssim 10^{26} \text{ poise} \qquad (5)$$

An anomalous mantle mass supplied to the crust
under tectonically active areas becomes segregated in the asthenosphere. It emerges rapidly
in the layer and it meets on its way the base of
a more viscous lithosphere. Here the rate of
its emergence sharply diminishes, and the
anomalous mantle material accumulates under the
lithosphere. As a result an isostatic uplift
originates with a value of

$$\zeta = \frac{\rho m - \rho}{\rho m} h \qquad (6)$$

where $\rho m = 3.35$ Mg/m^3 the density of the normal
mantle, ρ = the density of the anomalous mantle,
h = the thickness of its layer. For an isostatic compensation of an uplift $\zeta = 2.5$ km high
by a layer of an anomalous mantle h = 30 km
thick its density should be $\rho \approx 3.07$ Mg/m^3, i.e.,
280 kg/m^3 less than in a normal asthenosphere.

The pressure in an anomalous mantle under an
uplift is higher than the pressure at the same
depth in neighboring areas, where an anomalous
mantle is absent by a value of $\Delta p \sim \rho g \zeta$. For
uplifts several kilometers high $\Delta p \sim 100$ MPa. The
viscosity of an anomalous mantle is very low
($\eta \lesssim 10^{17}$ poise). Under an excessive pressure, it
should rapidly spread in a horizontal direction
along the base of the lithosphere.

Let us assume at first that the base of the
lithosphere is flat. In such a case with a
constant viscosity of the anomalous mantle the
time of its spreading τ for a distance L from
the area of ascent can be estimated as

$$\tau \sim \frac{3 \eta L^2}{(\rho m - \rho) g h^3} \qquad (7)$$

Let us take here $\eta \sim 10^{17}$ poise, h=20 km,

L=1000 km, $\rho m - \rho = 250$ kg/m^3. In such a case
$\tau \sim 50$ ka. This time is quite small as compared
with the time of processes in tectonically
active areas that is in the order of <11-10 Ma.

Along the base of the lithosphere the light
material can penetrate also into the stable
areas with an asthenosphere of normal viscosity.
In this case the main resistance to the spreading is rendered by the expulsed substance of the
asthenosphere. If the viscosity of the latter
at the lithosphere base is $\eta \sim 10^{19-20}$ poise, an
anomalous mantle layer ~ 10 km thick during a
period of several million years will penetrate
under a stable area to a distance from a few
hundreds to several thousands kilometers.
During a longer time period the anomalous mantle
layer of such a thickness cannot spread because
as a result of cooling at the expense of the
heat flow into the overlying lithosphere and the
underlying colder asthenosphere, its temperature
drops and the viscosity sharply increases.
After this the flow of the anomalous mantle
virtually stops.

In this way, the anomalous mantle can spread
along the base of the lithosphere over a large
distance. This, however, requires the viscosity
of lithosphere layers located over this boundary
to be sufficiently high. Otherwise, a convective instability arises and a lighter anomalous
mantle begins to get injected into the lithosphere.

The time of the convective instability evolution at the base of the lithosphere can be estimated by standard formulas describing this
phenomenon (Artyushkov, 1971b). The instability
has no time to progress during several million
years if the viscosity of the mantle over the
lithosphere base is $\gtrsim 10^{22}$ poise. Such a value
is much smaller than the average viscosity of
the lithosphere $\eta \gtrsim 10^{26}$ poise (Artyushkov,
1973).

The spreading of the light material over
large distances takes place only provided a drop
in viscosity at the lithosphere base amounts to

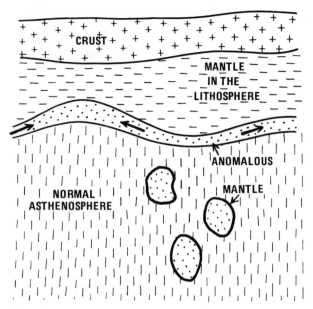

Fig. 3. Ascent of an anomalous mantle from the asthenosphere and its spreading along the base of the asthenosphere.

two orders of magnitude or more. This jump in viscosity should take place at a distance smaller than the thickness of the spreading layer; in other words, it should take place at a depth interval not exceeding several kilometers.

One cannot say a priori whether such a viscosity jump actually exists or not. It can originate, for instance, with a relatively small change in the activation energy at a certain temperature. For the time being, there are no data available that pertain to such a change in the activation energy. It should be noted, however, that the concepts of lithospheric plates drifting over large distances over the asthenosphere include an assumption of a pronounced difference in the viscosity of these layers. As demonstrated below, characteristic features of vertical crustal movements indicate that the spreading of the light material along the lithosphere base with a sharp jump in viscosity is actually taking place.

The thickness of the lithosphere is not uniform. At shallow depths the viscosity of the mantle is determined mainly by temperature. The latter varies greatly in a horizontal direction. Therefore, the thickness of the lithosphere d varies also. It rises under the cold platform areas and sharply drops under intensely heated tectonically active areas. Under platforms $d{\sim}100$ km, while under mountain structures it is $d{\sim}20\text{-}30$ km (Walcott, 1970; Balavadze and Tuliani, 1974). On platforms the mantle temperature is also non-uniform, which means that here there should also be variations in the thickness of the lithosphere.

Let us discuss the consequences of spreading

of light material along the base of a lithosphere with a variable thickness (Yanshin et al., 1977; Artyushkov et al., 1978). In the areas of a higher position of the lithosphere base (Fig. 3), the light material accumulates displacing the less heated and denser asthenosphere substance previously located here. In this way, the areas of a higher position of the lithosphere base play the role of "traps" for the light material. They must exist in places where the mantle temperature is higher, which decreases the thickness of the lithosphere. Conversely, areas with a lower mantle temperature correspond to a lower position of the lithosphere base. In its movement the light material tends to avoid them. Such sunken portions of the lithosphere base play the role of "antitraps." The flow of light material is accelerated in its approach to the traps and slows down in coming nearer to the sunken portions of the lithosphere base.

Let us see how traps are originating at the base of the lithosphere. Apparently, the entire continental crust, or its main part, at some time went through a stage of an orogenic evolution. The latter is characterized by an intense magmatic activity and, consequently, by high temperatures of the crust and mantle and a decreased thickness of the lithosphere. Under a mountain structure there usually is a large trap. After the end of the orogenic stage a large uplift on the surface of the earth becomes peneplained rather rapidly. And yet, the temperature inhomogeneity at depth remains present for a very long time. The cooling time of a layer of thickness h is

$$t \sim h^2/\kappa \tag{8}$$

where κ is the thermal diffusivity of the layer. The cooling of a crust 40 km thick with a thermal diffusivity $\kappa{\sim}10^{-6}$ m^2/s takes place during a time period $t{\sim}50$ Ma. In the top layers of the mantle the thermal diffusivity drops sharply to values $\kappa{\sim}2\text{-}3{\times}10^{-7}$ m^2/s. Therefore, its cooling to a depth of $h{\sim}100$ km corresponding to the average position of the lithosphere base in platform areas takes place during 300-500 Ma.

Platform Shields

During the cooling period of the material in the trap area under a former mountain structure, the density of the top layers of the mantle remains reduced. For this reason the crust over the trap will be isostatically uplifted with respect to the surrounding territory – in other words, here is a structure of the type of a crystalline shield. In the course of time the temperature in the top layers of the mantle gradually drops while the height of the trap diminishes. Simultaneously, the density of the mantle increases and there is a lesser height of

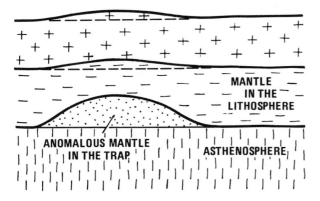

Fig. 4. Lithosphere uplift in a crystalline shield area at the entrapment of an anomalous mantle by a trap.

the territory with respect to the surrounding areas with a cold lithosphere. After 200-300 Ma, as it follows from (8), the top layer of the earth is cooled having a thickness of h~60-70 km. With a lithosphere thickness in adjacent stable areas of d~100 km, this corresponds to a height of the trap approaching to several tens of kilometers.

As stated before, the rise of light, heated material to the upper mantle takes place about once in 200 Ma, perhaps more often. For this reason if a trap formed during the previous mountain-building epoch appears to be near the ascent area of an anomalous mantle, it entraps a new portion of this material and the lithosphere over the trap experiences a new uplift (Fig. 4). The height of a trap under shields comes to several tens of kilometers - of the same size order as the thickness of its overlying mantle layer in the lithosphere. Therefore, the trapping of new portions of hot material results in a substantial heating of the lithosphere and a decrease in its thickness by ~10-20 km. During the next epoch of ascent of heated light material from the mantle, the process can be repeated if the trap is not too far away from the ascending flow in the upper mantle. Thus, having been formed once, the trap can be self-maintaining by entrapment of progressively newer portions of an anomalous mantle.

The magnitude of a territory uplifted over a trap is determined by the amount and temperature of the light material entrapped. If the trap is sufficiently far from the ascent area of the anomalous mantle, the supply of this material is relatively small and the new material has time to cool by flowing along the lithosphere base. In this case a portion of the light material can be intercepted on its way from the source by other traps. Under such conditions a minor uplift of the type of a crystalline shield is being maintained on the surface of the earth at the expense of a revival of ascending movements, while the anomalous mantle does not penetrate to

the crust remaining located in the lithosphere under the mantle.

At a relatively low temperature of the anomalous mantle in a trap, the velocities of longitudinal waves in it can be rather high: $V_p \sim 7.9-8.0$ km/s and perhaps higher.

With a consecutive multiple supply of relatively small volumes of a light material into the trap, crystalline shields like the Baltic or Canadian shields are maintained and have a long life. A periodic supply to the traps under the shields of an intensely heated material agrees poorly at first sight, with the low heat flow on shields. Lower heat flows in these areas are associated, however, first with the denudation of a substantial portion of the granitic-sedimentary layer with which the major part of the radioactive heat release in the crust is associated. In the trap the light material is separated from the crust by a layer of a normal mantle in the lithosphere ~20 km thick. The thermal diffusivity of the mantle in this area is very low: $\kappa \sim 2-3 \times 10^{-7}$ m^2/s. Consequently, even a sufficiently heated mantle in a trap cannot transmit to the surface any great amount of heat in the form of a conductive heat flow.

However, the mantle under the shields from time to time really experiences a substantial overheating. This is indicated by periodic magmatic phenomena on shields. Thus, on the Kola Peninsula, a part of the Baltic shield, where conditions typical for a shield exist already for over 1500 Ma, an injection of large intrusions took place in the Devonian.

As a result of its long-lasting high position, the crust on the shields is subjected to an intense denudation. No noticeable decrease of its thickness is, however, recorded. This can be due to a periodic melting out of a new crustal material from the mantle in a trap that joins the crust, thus increasing its thickness.

Areas of Tectonic Activization

If a structure of the crystalline shield type with a large trap is located near the ascent area from the depth of an anomalous mantle, it entraps it in a highly heated state and in great amounts, which is immediately accompanied by the formation of a substantial isostatic uplift. Simultaneously, an intense heating begins of the overlying mantle layer in the lithosphere, which greatly diminishes it viscosity. When the latter drops to values $\eta \sim 10^{22}-10^{23}$ poise, anomalous mantle can - as a consequence of convective instability - comparatively rapidly, within 1-10 Ma, expel the heavier mantle material overlying it. As a result, the anomalous mantle reaches the crust, while the mantle material previously located in the lithosphere moves downwards and sideways gradually being heated and acquiring temperature characteristics of the asthenosphere.

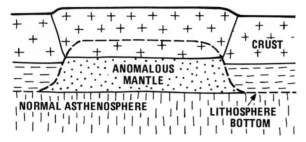

Fig. 5. Structure of the earth's crust and mantle under tectonically activated areas.

When it extends up to the crust, the trap reaches a great height and becomes entirely filled by a light material (Fig. 5). This leads to the formation of a high uplift that is isostatically compensated by a layer of light material. A typical example is the wide zone of uplifts in the western part of North America (Fig. 2). The thickness of the crust there does not exceed values normal for platform areas, despite the fact that the absolute height of the territory in its average is 2-2.5 km. This means that virtually the entire uplift is compensated by an anomalous mantle layer. In a number of regions the thickness of the crust there is reduced to 30 km and less, with a corresponding increase in the thickness of the anomalous mantle layer.

The compensation of the anomalous mantle uplift is also characteristic for the area around Baikal (Krylov et al., 1970, Puzyrev et al., 1974). There an anomalous mantle is located under a crust with normal platform thickness. In the Northern Tien Shan the crust is about 50 km thick - in other words, it is nearly the same as under the adjacent plains (45 km) (Krestnikov and Nersesov, 1962). The average height of the relief there comes to about 4 km. Consequently, the main portion of the uplift is fully compensated by a decrease in the mantle density under the crust. Such a decrease is also indicated by the seismic data (Vinnik and Lukk, 1975; Vinnik, 1976).

A number of tectonically active areas passed through a stage of platform evolution with the formation of shields in place of former mountain structures. The elevation of the shields gradually diminishes with time, and sometimes shields are covered by a thin sedimentary cover. Later on, during the activization epoch of tectonic movements, some shields are subjected to intense uplifts, becoming transformed into areas of tectonic activity.

So, for instance, folded mountain structures existed in the place of the Tien Shan during the Paleozoic, which completed their evolution about 220 Ma ago (Tectonic Map of Eurasia, 1966). After peneplanization of the relief, a shield was formed which by the Oligocene gradually sank and became locally overlain by sediments with a thickness up to several hundred meters. During

the Oligocene (25 Ma ago), an uplift of the region began - minor at first, but gradually intensifying with time. During the Pliocene and Quaternary time as a result of predominantly vertical movements, a high mountainous relief has been formed (Krestnikov, 1962). In this way, after the end of the Paleozoic orogenesis, a trap existed under the former mountains in the area of the present Tien Shan, which was filled by light material during the next epoch of its rapid ascent from the core-mantle boundary. Such traps existed also under other recent areas of tectonic activity. Renewed mountain building in areas of previous mountain building had been aptly called "Deutero-orogenesis" - secondary orogenesis (Bogolepov, 1968).

At the stage of expulsion by the light material of colder mantle substance from the lithosphere in the evolutionary process of a convective instability, the distribution of its injections is very complicated. This determines a differentiated character of the originating crustal movements. In certain cases the uplifts may also include deeply sunken cold regions with a much greater lithospheric thickenss. If the lithosphere base happens to be below the base of the light material layer under the adjacent mountain areas, such areas are not involved in the uplift. The Ferghana and Djungaria depressions belong, apparently, to structures of this type. Mountain building is accompanied by their rapid isostatic sinking because of an accumulation of a great amount of sediments washed down from the adjacent mountain ridges. These structures become intermountain and piedmont depressions.

As mentioned earlier, currently popular concepts explain the formation of mountain structures on continents by an increase in the thickness of the crust that resulted from horizontal compression of the lithosphere by certain forces external to the mountain areas.

Let us first discuss the formation of mountains under a more or less uniform compression of the platform lithosphere over a certain area. A characteristic feature in the compression deformation of a layer including a set of thinner sublayers with different geological properties is the formation of folds. If a continental crust ~40 km thick is compressed to a thickness of 60 km, the previously horizontal layers over the entire territory of the compressed area must be crushed into steep folds with angles of dip ~50-60°. However, it is known that such structures associated with the period of tectonic activization in mountainous areas are virtually absent. Only at the boundary zones between mountain systems and the adjacent plains local compression zones are originating which are not exceeding a few kilometers in width.

So the Tien Shan, for instance, has been formed as a folded area during the Paleozoic. During the latest mountain-building phase, deformations of the basement and of the sedimentary cover originated here in connection with a non-uniform uplift of the adjacent portions of the crust. The formation of mountain structures in the Baikal area also has not been accompanied by folding. On the contrary, tensile stresses exist in the latter region so that an uplift of the crust can undoubtedly not be associated with its compression.

Another mechanism associated with the formation of tectonically activated areas with horizontal movements is the underthrusting of one plate of the continental lithosphere under another. If both plates belong to platforms, then following underthrusting the sequence of the crust and mantle in the lithosphere should become repeated twice, and include the crust, the cold mantle in the lithosphere, one more crust and again a cold mantle in the lithosphere. No structure of such a type has ever been found.

Let us assume now that a cold platform plate is being thrust under another plate, which has already been heated previously. In that case the lithosphere in the upper plate does not contain any cold mantle layer and the crust in the underthrust plate proves to be located directly under the crust of the areas under which the thrusting is taking place. However, according to deep seismic sounding data, there is no place where a crust of double thickness could be established under areas of tectonic activity. Under Tien Shan, the Baikal area and in the western part of the USA, its thickness is approximately the same as under adjacent platforms or is even slightly smaller. Consequently, there has been no underthrusting of one portion of continental crust under another in these areas of tectonic activity.

Therefore, mountain structures within lithospheric plates in areas of tectonic activization do not originate by horizontal compression, but as the result of vertical movements. Their formation is associated with a supply to the crust of an intensely heated anomalous mantle with a density lower than in a normal mantle under a crust of platform areas. An increase in the relief height leads to the appearance of great additional stresses ~100-200 MPa (Artyushkov, 1973). These stresses can be both compressive, as in the marginal parts of Tien Shan, for instance, and tensile, as in the Baikal rift zone (Balakina et al., 1972). Associated with them is a high seismicity of many tectonically active areas.

Sedimentary Basins in Platform Areas

Sedimentary basins on platforms are formed in place of former shields. As pointed out earlier, the latter exist in an uplifted position with respect to the adjacent areas owing to a periodic supply of light material into the traps at the lower boundary of the lithosphere. If, however, at a certain time, the trap proves to be very far from the source of light material, it does not entrap light material at all or entraps it only in small quantities. In that case, the uplift of the crust and the heating of the lithosphere will be less pronounced or not take place at all. Because of temperature decrease in the top layers of the mantle, the amplitude of the trap decreases. Consequently, next time even if the light material is trapped, the amount is smaller and proves to be insufficient to maintain a high position of the earth's surface. After that, a progressive decrease in the height of the trap begins, and is followed by an increase in the lithosphere thickness, accompanied by an increase in the density of the mantle in it, and sinking of the area.

When the base of the lithosphere sinks into the place of a former trap, the light heated material begins to avoid its downwarped portion. As a result, such structures remain cold all the time and also prove to be self-supporting.

In the formation of a sedimentary basin of depth h, the load upon the lithosphere increases by a value $\rho_s gh$. Simultaneously, when the lithosphere becomes downwarped, an asthenosphere matter with a ρ_m density becomes squeezed from underneath, which decreases the load on the compensation surface in the asthenosphere by a $\rho_m gh$ value. The summary change in the load

$$\Delta p = (\rho_m - \rho_s)gh \qquad (9)$$

is negative since $\rho_m > \rho_s$; the asthenosphere is always denser than the sediments. With $\rho_m = 3.35$ Mg/m^3, $\rho_s = 2.55$ Mg/m^3, and a depth of the sedimentary basin h=10 km, we find $\Delta p = -80$ MPa. If the sinking of the lithosphere would proceed without a simultaneous change in its density, a negative isostatic gravity anomaly of an order of 3 mm/s^2 would appear over a basin of such a depth.

Actually, isostatic anomalies both over the shields and over sedimentary basins do not exceed several tens of milligals. This means that along with a transformation of the shields into sedimentary basins a substantial increase in density is taking place in the lithosphere. More exactly, such an increase in density (compaction) causes sinking of the territory.

A most substantial effect is the compaction within the basaltic layer of the crust (see below) and within the trap at the lower boundary of the lithosphere. Let us designate a decrease in the thickness of the crust and of the anomalous mantle layer in a trap under a compaction as Δh_c and Δh_a, correspondingly. It is not difficult to show then by using the condition of isostatic equilibrium of the lithosphere that

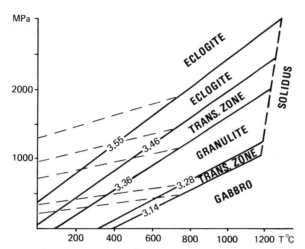

Fig. 6. Phase transformation diagram of a gabbro into a garnet granulite and an eclogite. Dashed lines indicate different variants of curve extrapolation into low temperature areas (Ito and Kennedy, 1971).

the depth of a sedimentary basin

$$h_s = \frac{\rho_m}{\rho_m - \rho_s} \; (\Delta h_c + \Delta h_a) \qquad (10)$$

It is $\rho_m/(\rho_m - \rho_s)$ times larger than the summary reduction in the thickness of the crust and layer of light material in the trap. With an average density of sediments

$$\rho_s \simeq 2.55 \text{ Mg/m}^3 \text{ and } \rho_m = 3.35 \text{ Mg/m}^3$$

$$h_s \simeq 4.2 \; (\Delta h_c + \Delta h_a) \qquad (11)$$

The formation of a sedimentary basin of such a depth by rock compaction in the crust and mantle can be explained in the following way. An increase in the density leads to compaction of the crust and mantle by the value $\Delta h_c + \Delta h_a$ and the formation of a depression of the same depth at the surface. The filling out of the depression by sediments increases the load upon the lithosphere and is accompanied by its isostatic subsidence to the value of (11). With the cooling of an anomalous mantle in the trap one can expect its compaction by a value up to $\Delta\rho \sim 200{-}300 \text{ kg/m}^3$. A decrease in the thickness of the rock layer Δh of an initial density ρ_o at a compaction by the value $\Delta\rho$ is

$$\Delta h = (h_1 - h_o) = \frac{\Delta\rho}{\rho_o} \; h_1 \qquad (12)$$

where h_1 is a final thickness of the layer. With $\Delta\rho \sim 200 \text{ kg/m}^3$, $\rho_o \simeq 3.15 \text{ Mg/m}^3$, and a height of the trap being 20–30 km $\Delta_a \sim 1.2{-}1.8$ km. The

depth of a sedimentary basin formed by this mechanism is, according to (11),

$$h_s \sim 5 - 8 \text{ km} \qquad (13)$$

Let us now consider density changes within the crust. According to present concepts, the lower, "basaltic" geophysical layer of the crust consists mostly of basic rocks. At low pressure or high temperature, their stable state corresponds to an association of minerals that forms gabbro with a density $\rho \sim 2.9{-}3.0 \text{ Mg/m}^3$. At higher pressure or a lower temperature an association becomes stable that corresponds to eclogite with a density $\rho \sim 3.4{-}3.6 \text{ Mg/m}^3$. Figure 6 shows a phase diagram of the gabbro-eclogite transition for an oceanic tholeite (Ito and Kennedy, 1972). The transition does not proceed along a sharp boundary but in a wide transitional zone where, with an increase in pressure, the density of the rock gradually increases along with a greater content of garnet. In this zone there exist garnet granulites and plagioclase eclogites. Experiments have been proceeding under high temperatures of >700–800°C only. Figure 6 shows different variants of curve extrapolation into an area of lower temperatures.

The inclination angle of the curves and the width of the transition zone differ for basalts of a different composition (Green and Ringwood, 1967). That is why one cannot definitely assert that the diagram on Figure 6 applies to basalts in the lower layer of the crust. Another point is that one cannot exclude the chance of their composition in different parts of the earth being varied, which means the phase diagrams would also be different. Nevertheless, it is possible to have a qualitative discussion on the basis of the above-mentioned diagram.

The temperature at the Moho under the platform is ~400–500°C (Lubimova and Smirnov, 1977). The pressure in the basaltic layer at an average depth from 20 to 40 km comes to 500–1100 MPa and the temperature in it is ~300–500°C. Under such conditions high-density rocks standing close to eclogite are stable. However, at a temperature of 300–500°C the transition develops extremely slowly. Its characteristic time under such conditions is ≥ 1000 Ma (Sobolev, 1978). The associated increase in crustal density extends over a very long time.

Seismic studies reveal under many platform areas a layer in the low part of the crust with a thickness of 10–15 km and velocities of longitudinal waves Vp~7.1–7.3 km/s. Such velocities are not characteristic for a basalt, but more so for a garnet granulite with a density of $\rho \sim 3.1{-}3.2 \text{ Mg/m}^3$. In a number of regions, under deep depressions particularly, even higher velocities Vp~7.4–7.6 km/s are observed in the lower parts of the crust which corresponds to a

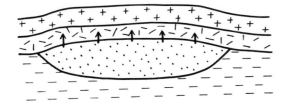

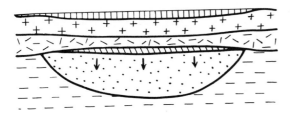

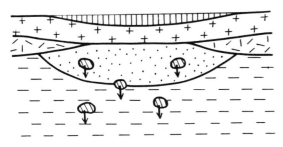

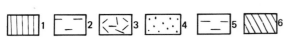

Fig. 7. Consecutive stages of processes leading
to the formation of an inland sea during a
supply of an anomalous mantle to the cold crust
of a continental platform: (a) uplift at the
moment of an anomalous mantle approach to the
crust, (b) sinking during a transformation of a
part of the basaltic layer into a denser garnet
granulite and eclogite, (c) tearing off of the
eclogitized basaltic layer and the formation of
a deep basin (1-sedimentary layer together with
a layer of water; 2-granitic layer, 3-basaltic
layer; 4-anomalous mantle; 5-normal mantle;
6-garnet granulites and eclogites).

garnet granulite with a density $\rho\sim3.3-3.4$ Mg/m^3.
Such a situation is characteristic, for instance, for the Ferghana and Kura depressions
(Volvovsky and Volvovsky, 1975; Sollogub,
1978).

A basalt compaction in a transition into a
garnet granulite by a value of $\Delta\rho\sim200-300$ kg/m^3
of a layer with final thickness of 10 to 15 km
thick causes, according to (12), a vertical
contraction of the crust by $\Delta h_c\sim0.7-1.5$ km which
in turn results in the formation of a sedimentary basin of a depth of

$$h_s \sim 6 \text{ km} \qquad (14)$$

A particularly substantial sinking is caused
by a transition of the lowest part of the basaltic layer of the crust into eclogite. In this
process $\Delta\rho\sim500-600$ kg/m^3. A transition in a
layer of finite thickness of 7 km produces a
crust contraction by 1.2-1.4 km. The depth of
sedimentary basin thus formed is

$$h_s \sim 5 - 6 \text{ km} \qquad (15)$$

and comes close to the thickness of the layer
that experienced a transition into eclogite.
The most typical velocities of longitudinal
waves in eclogite $Vp\sim8.1-8.2$ km/s come close to
a velocity in a peridotitic mantle located under
the crust. Therefore, it is very difficult to
distinguish below the crust eclogite from
peridotite by seismic methods. A total downwarping of a region during a transition from
shields to sedimentary basins can be produced by
any of the processes described, by two of them
or even by all three processes combined - a
mantle consolidation in the area of a former
trap, a transition of the lower part of the
basaltic layer into eclogite and of a part of
the overlying basalt into a garnet granulite.
Judging by greater velocities in the lower part
of the crust under certain shields, a compaction
of its lower layer begins already at this stage.
The sinking, however, is delayed owing to the
presence of light material in the trap.

As it follows from (13)-(15), in total these
processes are able to provide a subsidence of
the crust by 15-18 km, compensated by sedimentation.

Inland Seas

When very large masses of a heated light
material are supplied to the base of the lithosphere, this material cannot only fill traps,
but may also form a rather thick layer under the
subsident areas of the lithosphere base. By
heating the mantle in the lithosphere, it makes
it less viscous, expels the colder and denser
substance of the mantle from the lithosphere and
comes into direct contact with the basaltic
layer of the crust (Fig. 7a). In this process
the basalts in the lower parts of the crust
begin to get heated.

The rate of the gabbro-eclogite phase transition is very low at temperatures of 400-500°C,
that may be typical for the lower parts of the
crust on platforms, but sharply increases with a
rise in temperature - by about one order of
magnitude for each 100°. When the temperature
rises to ~800°C, the characteristic time of a
transition diminishes to ~1 Ma, while the
increase in the density at the transition is
still substantial (Sobolev, 1978). When such
temperatures are reached, a rapid - from a
geological point of view - transition of basalt

into a garnet granulite and then into eclogite takes place (Fig. 7b).

The formation of dense rocks at pressures characteristic for the basaltic layer of continental platforms is only possible at relatively low temperatures. For instance, as demonstrated by Figure 6, in an oceanic tholeite a great increase in density takes place only at temperatures $T \sim 800-900°C$. At higher temperatures $T \gtrsim 1000°C$ the phase transition, though developing very rapidly, does not produce any substantially greater density. For this reason, in mountain areas where a highly heated mantle is located under the crust, the lower crustal layers do not become heavier.

A rapid increase of density in a basaltic layer of the crust is in this way possible, only within a narrow interval of temperatures, when the phase transition is already taking place at a high rate but still results in a substantial compaction of the rock. Such conditions can exist upon penetration of the light material to the crust in cold platform areas and particularly under sedimentary basins with a thick lithosphere. Having cooled during the expulsion of the cold mantle from the lithosphere, the light material should approach the crust being heated by several hundred degrees less than under areas of tectonic activization. Blocks of a dense garnet granulite and eclogite formed at the places of contact with the crust of a light heated mantle, become torn off from the gravity field of the crust and sink in the anomalous mantle of low viscosity (Fig. 7). As a result the anomalous mantle comes into contact with new portions of the basaltic layer which are also undergoing a phase transition, and break off from the crust to sink in the mantle. Along with the sinking of a garnet granulite with an increasing pressure, it changes into an eclogite with a density $\rho \sim 3.5-3.6$ Mg/m^3, a rock denser than the normal mantle in the asthenosphere. Eclogite sinks into the asthenosphere and accumulates at its base.

As a result of the process described, a thick layer gradually becomes torn off from the crust corresponding to the entire basaltic layer or its lower part. The thinned crust experiences a rapid isostatic sinking with respect to the original position. The depth of the sea basin formed can be easily determined from the condition of isostatic equilibrium:

$$h_s = \frac{\rho_m - \rho_b}{\rho_m - \rho_o} h_b - \frac{\rho_m - \rho_a}{\rho_m - \rho_o} h_a \qquad (16)$$

Here ρ_b, h_b represent the initial density and thickness of the basaltic layer torn off from the crust, h_a the thickness of anomalous mantle layer under the crust, and ρ_o the density of the water. Let us assume in (16) the thickness of the eclogitized layer of basalt to be $h_b=20$ km

and its initial density $\rho_b=2.9$ Mg/m^3. Let the thickness of the anomalous mantle layer be $h_a=20$ km, and its density $\rho_a=3.25$ Mg/m^3 (owing to a relatively low temperature of $\sim 800°C$, the density of the anomalous mantle should be relatively close to the density of the normal mantle). In such a case, it follows from (16) that $h_s=3$ km. At a more substantial initial thickness of the basaltic layer $h_b=30$ km, which corresponds to a more drastic shrinking in the thickness of the consolidated crust, the depth of the basin comes to $h_b=4.9$ km. The sea basin is gradually filled out by sediments which increases the sinking depth of the crust. The latter increases also with further cooling and compaction of the anomalous mantle. The final depth of the sedimentary basin comes to:

$$h_s = \frac{\rho_m - \rho_b}{\rho_m - \rho_s} h_b \qquad (17)$$

With $\rho_s=2.55$ Mg/m^3, $h_b=20$ km, and $h_b=30$ km we compute from (17) $h_s=11$ km and $h_s=17$ km. It should be noted that these values refer only to the thickness of sediments accumulated after the destruction of the basaltic layer of the crust. The thickness of sediments formed before the beginning of this process is not included.

It is essential that the sinking of the region is always preceded by an uplift. It originates at the time when the light material approaches the base of the lithosphere and continues at the penetration stage of the light material to the crust.

Let us investigate the South Caspian basin as a typical example of an inland sea. The thickness of sediments here is 20-25 km (Beliaevsky, 1974). The lower 10 km of sediments accumulated at least since the Jurassic and possibly even since the Paleozoic, i.e., during >150 Ma, while the upper layer up to 15 km thick was deposited since the Middle Pliocene during the last 4-5 Ma (Tectonics of oil and gas areas, 1973). Before its latest subsidence, a platform sedimentary basin existed in the place of the Southern Caspian, i.e., the thickness of the continental crust here should be normal for platforms. At the present time it is reduced under the Southern Caspian to 10-12 km. Shortly before the beginning of the sinking an uplift has been taking place along its periphery (Volvovsky and Shlezinger, 1976), Middle Pliocene deposits overlie the subjacent horizons with a sharp unconformity. Consequently, the abyssal part of the South Caspian basin must have been most probably uplifted just before the recent subsidence.

The area described is characterized by an extremely intense heating of the top layers of the mantle (Ashirov et al., 1976). The layer of higher electrical conductivity under the South

Caspian basin rises to a depth of 40-60 km. On its periphery it is at a depth of 100-120 km, and sinks under the Scythian plate to 200-300 km. It is difficult to determine exactly the temperature at the Moho when the sediments are of a great thickness. It is estimated to be 600-1000°C.

In this manner, the area discussed is characterized by all the signs of a rapid sinking of the crust caused by a phase transition of basalt into eclogite: an initial platform stage of evolution a subsequent uplift of the territory, replaced by a sharp sinking, a strongly decreased thickness of consolidated crust and an intense heating of the mantle. In the abyssal part of Southern Caspian rocks with "basaltic" velocities $V_p \sim 6.8$-6.9 km/s are occurring with a thickness of 10-12 km. The "granitic" geophysical layer is absent here. This can be associated with metamorphic reactions in the "granitic" layer at a contact with the hot anomalous mantle, which substantially change the velocities of elastic waves.

The formation of inland seas according to the scheme described above differs in principle from a basification (Beloussov, 1968). Basification is a change in the composition of the crust caused by a supply of an ultrabasic material from the mantle. In the process of basification, the top layer of the crust becomes basic, while the lower becomes ultrabasic acquiring the density of the mantle or an even greater density.

In the above discussed process, there is no supply of ultrabasic material to the crust from the mantle. Instead the process is due to a heat supply to the crust from the mantle. In this process the average chemical composition of the crust — both of its top layer remaining at the surface and of the lower sinking into the mantle — remains unchanged. By an eclogitization of the lower layer of the crust we can explain the origin of a number of other depressions in the Alpine geosyncline belt of Eurasia: the Black Sea, the Pannonian basin, and deep basins of the Western Mediterranean (Artyushkov et al., 1978). The Pericaspian depression at the present time is completely filled by sediments and may serve as an example of an old basin of an inland sea type (Yanshin et al., 1977).

Thus, the ascent of a light heated material from the mantle may be a cause of intense uplift and sinking crustal movements depending upon whether it gets into previously heated or cold areas.

At the present time many authors regard inland seas as the results of an extension with the formation of an oceanic crust (Kropotkin, 1961; Biju-Duval et al., 1977). These concepts do not agree with geological and geophysical data (Lebedev et al., 1978; Artyushkov et al., 1978).

First of all, slightly deformed layers of platform sediments, usually of a great thickness, occur almost everywhere in inland seas under abyssal sediments. As a result of extension, they would have been immediately torn. In such cases abyssal sediments would be deposited directly on the basaltic crust.

Second, inland seas are usually more or less isometric in plan. An extension of isometric structures should have been accompanied by a compression of the crust around the basins - in areas of the same width as the width of the basins themselves. A contraction of horizontal scales here should have been of about 1.5 times or by half which, as it is known, leads to the formation of folds with angles of dip of $\geq 45°$, upthrusts and overthrusts. Such structures in the Alpine geosyncline belt of Eurasia near inland seas are found nearly everywhere. They originated, however, before the inland seas were formed, most of which cases were during the Pliocene-Quaternary time when the main folding had already been completed.

And finally, the process of extension and formation of the oceanic crust is accompanied by the formation of typical linear magnetic anomalies, according to which the very fact of extension is usually established. In the inland seas of the Mediterranean belt there are no such anomalies.

The Nature of an Inherited Character of Vertical Crustal Movements

As we have demonstrated, the same direction of vertical movements in many areas is associated with the existence of inhomogeneities at the lithosphere base over very long times. Uplifts of this boundary (traps) periodically entrap the light material while the areas overlying them experience ascending movements. The downwarps of the lithosphere base (antitraps) are usually avoided by the light material and the areas located above them remain cold and are subject to sinking.

Inhomogeneities in the lithosphere thickness are preserved during large horizontal displacements of lithospheric plates. Consequently, despite displacements of the latter for thousands of kilometers, the inheritance of vertical movements associated with these inhomogeneities is also preserved. The ascent from the depth of the light material takes place in a number of areas of the earth. If during the drifting of lithospheric plates the traps prove to be near ascending flows of the light material, they entrap it again while the downwarps of the lithosphere base are still avoided by this material. In this way the unidirectional character and lengthy duration of vertical crustal movements in many areas do not contradict the concept of continental drift.

The character of vertical movements is lost

when the structure of inhomogeneities at the lithosphere base changes. In this way, if a trap on a platform does not receive a supply of light material and its accompanying heat for a long time, an antitrap is being formed in its place. As a result a crystalline shield subjected to an uplift becomes transformed into a sinking sedimentary basin.

When an anomalous mantle is being supplied under the crust below inland seas, the previously existing cold mantle in the lithosphere under sedimentary basins is being expelled. Along with it disappear the inhomogeneities in the lithosphere which results for a time in a stop of the evolution of the structures of a higher order within the sedimentary basin.

References

Alexeev, A. S., and V. Z. Ryaboy, New structural model of the upper mantle of the earth, Priroda, 7, 64-77, 1976 (in Russian).

Artyushkov, E. V., Gravity convection in the earth's interior, Izv. Akad. Nauk SSSR, Ser. Fiz. Zemli, 9, 3-17, 1968 (in Russian).

Artyushkov, E. V., Density differentiation on the core-mantle interface and gravity convection, Phys. Earth Planet. Interiors, 2, 318-325, 1970.

Artyushkov, E. V., Rheological properties of the crust and upper mantle according to data on isostatic movements, J. Geophys. Res., 76, 1376-1390, 1971a.

Artyushkov, E. V., Convective instability in geotectonics, J. Geophys. Res., 76, 1397-1415, 1971b.

Artyushkov, E. V., Stresses in the lithosphere caused by crustal thickness inhomogeneities, J. Geophys. Res., 78, 7675-7708, 1973.

Artyushkov, E. V., Geodynamics, M. NAUKA, 328p., 1979.

Ashirov, T., V. G. Dubrovsky, and Ya. B. Smirnov, Geothermal and geoelectric investigations in the South Caspian basin and the nature of the high conductivity layer, Dokl. AN SSSR, 226, 1976 (in Russian).

Balakina, L. M., A. V. Vedenskaia, N. V. Golubeva, L. A. Misharina, and E. I. Shirokova, Field of elastic stresses of the earth and the mechanism of earthquake foci, M., NAUKA, 1972 (in Russian).

Balavadze, B. K., and L. I. Tuliani, On the inhomogeneity in the crustal structure of the Caucasus region, Dokl. AN SSSR, 217, 1379-1382, 1974 (in Russian).

Beliaevsky, N. A., The earth's crust within the territory of the USSR, M., NAUKA, 280p, 1974 (in Russian).

Beloussov, V. V., Basic problems in geotectonics, McGraw Hill, no. 4, 809p, 1962.

Beloussov, V. V., The crust and upper mantle of the continents, M., NAUKA, 1966, (in Russian).

Beloussov, V. V., The crust and upper mantle of the oceans, M., NAUKA, 1968 (in Russian).

Biju-Duval, B., J. Dercourt, and X. Le Pichon, From the Tethys Ocean to the Mediterranean Sea: A plate tectonic model of the evolution of the Western Alpine system in Structural history of the Mediterranean basins, Paris, Editions Technip., 143-164, 1977.

Bogolepov, I. V., On two types of orogenesis, Geologia i Geofizika, 8, 1968 (in Russian).

Bulmasov, A. P., V. P. Gornostaev, M. M. Mandelbaum, V. P. Pospeev, and K. A. Savinsky, Deep magneto-telluric soundings in the Baikal area in Baikalsky rift, M., NAUKA, 140-147, 1968 (in Russian).

Cook, K. L., Rift system in the basin and range province in The world rift system, Geol. Surv. Can. Paper 66-14, 246-279, 1967.

Dewey, J. F., and J. M. Bird, Mountain belts and the new global tectonics, J. Geophys. Res, 75, 2625-2647, 1970.

Dubrovsky, V. A., and V. A. Pankov, On the composition of the eath's core, Izv. AN SSSR, Ser. Fiz. Zemli, 7, 1972 (in Russian).

Florensov, N. A., Mesozoic and Cenozoic basins of the Baikal area, L., Izd-vo AN SSSR, 257p, 1960 (in Russian).

Foucher, J. P., Mechanical model of subsidence in the Paris basin in Sedimentary basins of the continental margin and craton, Durham, 13, 1976.

Grachev, A. F., Rift zones of the earth, L., NEDRA, 248p, 1977 (in Russian).

Green, D. H. and A. E. Ringwood, An experimental investigation of the gabbro-eclogite transformation and its petrological application, Geochim. Cosmochim. Acta, 31, 767-833, 1967.

Holmes, A., Principles of physical geology, 2nd ed., New York, 1965.

Ito, K., and G. C. Kennedy, An experimental study of the basalt-garnet granulite-eclogite transition in The Structure and physical properties of the earth's crust, Geophys. Monogr. v. 14 (J. G. Heacock, ed.), 1971.

Krestnikov, V. N., Evolution history of oscillating movements on the Pamirs and in adjacent parts of Asia.M., Izd-vo AN SSSR, 1962 (in Russian).

Krestnikov, V. N., and I. L. Nersesov, Tectonic structure of the Pamirs and Tien Shan and its relation to the Moho discontinuity, Sov. Geologia, 11, 1962 (in Russian).

Kropótkin, P. N., Paleomagnetism, paleoclimates and the problem of large-scale horizontal crustal movements, Sov. Geologia, 5, 1961 (in Russian).

Krylov, S. V., B. P. Mishenkin, G. V. Krupskaia, G. V. Petrik, and T. A. Yanushevich, Crustal structure according to the deep seismic sounding profile across the Baikal rift zone, Geologia i Geofizika, 1, 1970 (in Russian).

Lebedev, L. I., Ya. P. Malovitzky, M. V. Muratov, A. E. Shlezinger, and A. L. Yanshin, Comparative tectonic analysis of sedimentary covers in abyssal basins of the Mediterranean belt, _Abstracts of Reports_ in _Tectonics of the Mediterranean belt_, M., 1978 (in Russian).

Levin, B. Yu., Origin of the earth, _Izv. AN SSSR. Ser. Fiz. Zemli_, 7, 1972 (in Russian).

Lubimova, E. L. and Ya. B. Smirnov, eds., Map of mantle heat flow and Moho-temperature iso-lines, _Geophys. Res. Bull. of India_, 1-2, 1977.

Lyustikh, E. N., On the thalassogenesis hypo-theses and on crustal blocks, _Izv. AN SSSR, Ser. Geophys._, 11, 1959 (in Russian).

Magnitzky, V. A., The internal structure and physics of the earth, Washington, _NASA technical translations_, 447p, 1967.

McKenzie, D. P. Active tectonics of the Medi-terranean region, _Geophys. J.R. Astr. Soc._, 30, 109-185, 1972.

Nowroozi, A. A., Seismo-tectonics of the Persian plateau, Eastern Turkey, Caucasus and Hindu-Kush region, _Bull. Seismol. Soc. Am._, 61, 317-341, 1971.

Puzyrev, N. N., M. M. Mandelbaum, S. V. Krylov, B. I. Mishenkin, G. V. Krupskaia, and G. V. Petrik, Deep structure of the Baikal rift according to explosion seismology data, _Geol. i. Geophys._, 5, 1974 (in Russian).

Schmidt, O. Yu. Selected papers, geophysics and cosmogony, _Izd-vo AN SSSR_, 1960 (in Russian).

Smirnov, Ya. B., ed., Map of deep temperatures on the territory of the USSR and in adjacent regions, M., GUK, 1977 (in Russian).

Sobolev, S. V., Models of the lower part of the crust on continents with due consideration of the gabbro-eclogite phase transition in _Petrological problems of the crust and upper mantle_, Novosibirsk, NAUKA, 347-355, 1978 (in Russian).

Sollogub, V. B., _Deep structure of Eastern Europe_, Kiev, NAUKOVA DUMKA, 1978 (in Rus-sian).

Talwani, M., X. Le Pichon, and M. Ewing, Crustal structure of the mid-ocean ridge 2. computed model from gravity and seismic refraction data, _J. Geophys. Res._, 70, 1965.

Tectonic map of Eurasia, Yanshin, A. L., ed., M., NAUKA, 1966 (in Russian).

Tectonics of oil and gas areas in the south of the USSR, Trudy, _VNIGNI_, fasc. 141, 1973 (in Russian).

Van Bemmelen, R. W., The undation theory of the development of the earth's crust in _Proc. 16 Int. Geol. Congress_, Washington, D.C., 2, 965-982, 1933.

Vanyan, A. L., and E. P. Harin, Deep magneto-variational soundings in the Baikal area in _Regional geophysical investigations in Siberia_, Novosibirsk, NAUKA, 184-193, 1967 (in Russian).

Vinnik, L. P., _Studies on the earth's mantle by seismic methods_, M., NAUKA, 198p, 1976, (in Russian).

Vinnik, L. P., and A. A. Lukk, Horizontal inhomo-geneities of the upper mantle in the areas of platform activization of Central Asia, _Izv. AN SSSR, Ser. Phys. Zemli_, 7, 1975 (in Russian).

Volvovsky, B. S., and I. S. Volvovsky, Crustal sections of the territory of the USSR accord-ing to the data of deep seismic soundings. M., _Sovetskoe radio_, 264p, 1975 (in Russian).

Volvovsky, I. S., and A. E. Shlezinger, The position of the Black Sea and South Caspian basins in the structure of the crust in _The crust of the continental margins and inland seas_, M., NAUKA, 1977 (in Russian).

Walcott, R. I., Flexural rigidity, thickness and viscosity of the lithosphere, _J. Geophys. Res._, 75, 3941-3954, 1970.

Yanshin, A. L., E. V. Artyushkov, R. G. Goretzky, L. G. Kiriukhin, R. B. Sapozhnikov, and A. E. Shlezinger, Comparative characteristics of the origin and evolution history of the Turanian plate and the Pericaspian basin, _Abstracts of Reports_ in _Problems of tectonics of the territory of the USSR and distribution of minerals_, M., Interdepartmental tectonic committee, 18-21, 1977 (in Russian).

Yanshin, A. L., E. V. Artyushkov, and A. E. Shlezinger, Main types of large structures of lithospheric plates and possible mechanisms of their formation, _Dokl. AN SSSR_, 234, 1977.

DYNAMICS OF THE RUSSIAN AND WEST SIBERIAN PLATFORMS

A. L. Aleinikov, O. V. Bellavin, Yu.P. Bulashevich, I. F. Tavrin

Uralian Scientific Centre, Swerdlovsk, USSR

E. M. Maksimov, M.Ya. Rudkevich

West Siberian Petroleum Institute, Ul. Volodarskogo, Tjumen, USSR

V. D. Nalivkin, N. V. Shablinskaya

VNIGRI (Union Petrol. Scientific Geological Research Institute)
Liteinii pr. 39, Leningrad, D-104, USSR

V. S. Surkov

SNIIGGIMS (Siberian Scientific Geological and Geolophysical and Mineral
Resource Institute), Nowosibirsk, Krasny pr. 67, USSR

Abstract. On the Russian, West Siberian and some other platforms, sedimentation began with the formation of major grabens (aulacogens). Structures of this type are apparently most common during the initial stages of platform development. The aulacogen and geosynclinal downwarping is associated with epochs of major transgressions while tectonic inversion of aulacogens and orogeny are associated with regressions. Different size platform structures of ancient and young platforms suggest an increase in growth rate during the beginning of transgressions and decrease in growth rate during regressions. All this implies the existence of uniform periodical tectonic processes that may be operative over vast areas of the globe. Average growth rates of platform structures are very low, i.e., 0.15-5 m per Ma. On the ancient Russian platform, such rates were smaller than those on the young West Siberian platform. The duration of structural growth can be rather long. On the young West Siberian as well as on the more ancient Russian platform, structures continue to grow and today many of them are reflected in the topography. This suggests that deep-seated loci of tectonic movements are persistent both in time and with respect to platforms. Major fractures have an even longer duration, often extending over 25 percent of the earth's age. Fractures on the platforms discussed in this paper are most widespread near the top of the basement. They decrease in number in the upper parts of the sedimentary cover and apparently in the lower crust. Deep seismic soundings clearly show that the loci of tectonic movements responsible for the formation of the major and biggest structures on both ancient and young platforms lie below the earth's crust. Their evolution should be concurrent with those of lithosphere plates. The presence of steep fractures is another argument against permanent and considerable horizontal displacements in the earth's crust.

1. Dynamics of the Development of the Russian Platform Structures

Eleven major sedimentary basins and six big uplifts that separate them occur on the Russian platform. Except for the Baltic and Ukrainian shields all of them are covered by a continuous but not coeval sedimentary cover (Fig. 1). The accumulation of unmetamorphosed sediments began in the late Proterozoic (Riphean time) about 1400 Ma ago. These sediments rest on a highly metamorphosed early Proterozoic basement. In the northeastern part of the platform beyond the Timan Ridge the basement is much younger, i.e., late Proterozoic. There the earliest rocks of the platform cover are of Ordovician age (510 Ma).

Types of Tectonic Structures

The authors subdivide all the structures into five groups according to their size (Table 1).
The following types of the biggest and major structures are recognized within the Russian platform:
1. Intraplatform circular structures (syneclises, anteclises, depressions, arches) caused by flexure in the earth's crust.

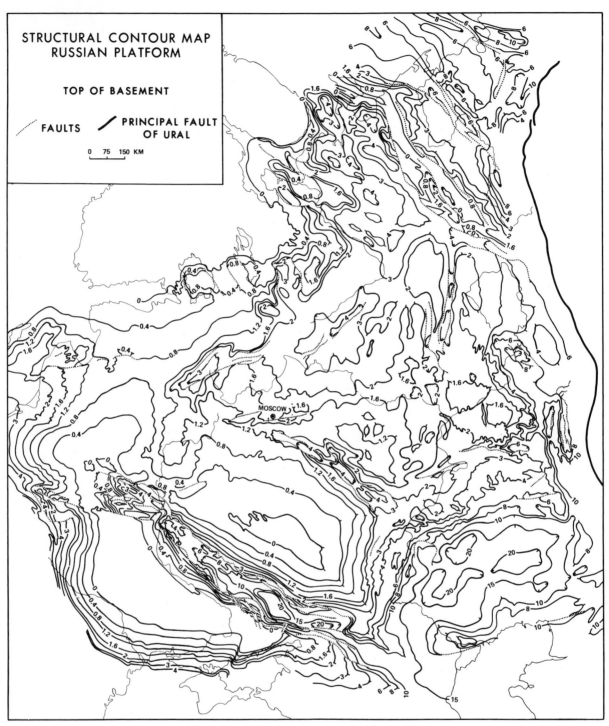

Fig. 1. Structure contour map of the Russian platform for top of the basement (after Bronguleev et al., 1975).

2. Intraplatform highly elongate large grabens (aulacogens) due mainly to fault movements.
3. Embayments at the platform margins and involving subsidence of actively subsiding geosynclines (pericratonic embayments).

These different types of basinal structures are very often superimposed. Syneclises, depressions, pericratonic embayments and aulacogens are all grouped together by the term "sedimentary basin."

Table 1

SIZE	SUBDIVISION OF CRATON	BIGGEST	MAJOR	MEDIUM	LOCAL
(AREA) OF STRUCTURES IN 10^3 km^2	>600 - 1000	60 - 100 TO 600 - 1000	6 - 10 TO 60 - 100	0.2 TO 6 - 10	< 0.2
NEGATIVE STRUCTURES	PLATFORM	SYNECLISES, PERICRATONIC EMBAYMENTS,	DEPRESSIONS (BASINS), MINOR GRABENS-AULACOGENS	TROUGHS, ELONGATE TROUGHS	LOCAL DEPRESSIONS
POSITIVE STRUCTURES	SHIELDS	ANTECLISES	ARCHES, MAJOR SWELLS	SWELLS, DOMES	LOCAL UPLIFTS

Aulacogens predominate among the sedimentary basins of the Russian platform. There are seven aulacogens, six inner syneclises and four pericratonic embayments. It is noteworthy that volumes of their sedimentary fill show an opposite trend. About 4 million km^3, 3 million km^3, and 2.5 million km^3 are the volumes for pericratons, inner syneclises, and aulacogens, respectively.

The Precaspian syneclise is a rather rare type of the biggest structures. It occupies the southeastern corner of the platform. The depth of its basement is very great (20-22 km). Its flanks are complicated by a system of step-like faults but unlike aulacogens, it is a circular basin. The granitic layer is thin and in places absent. The Precaspian syneclise is similar to the Mesozoic and Tertiary basin of the Gulf of Mexico; however, the downwarping in the Precaspian synclise took place mainly during the close of the Paleozoic and in the early Triassic. It contains about 40 percent of all the sedimentary rocks accumulated on the Russian platform. The deep structure of the Precaspian basin is poorly known; therefore, figures are not given together with average estimates presented below.

Medium and local structures, as well as major ones, are subdivided into those that resulted from flexure in the earth's crust and those that are related to fractures. The latter are elongate and bounded by fault scarps and their flanks dip about 20° to 50°. Fault structures of different size usually occur together and local structures complicate the larger ones.

A very interesting graben type about 1 km wide and several tens of meters to 120 m deep and about 250 km long is known in the eastern part of the Russian platform. They diverge forming a slightly spread-out fan and are spaced 20-30 km apart. Their subsidence is over short time spans (several Ma) and not always the same time, i.e., the close of the Middle Devonian and

early Upper Devonian (Khat'yanov, 1971).

The sedimentary mantle of the Russian platform is disturbed by many fractures that occur on quite different scales. The most extensive fractures are 250-400 km long with a throw of 1 km at basement level. They form the boundaries of grabens or aulacogens. Less extensive fractures are related to smaller structures or else they do form step-like systems. Their throw amounts to a few tens of meters. Fractures are most abundant within the Precambrian formations and near the basement. Late Permian and Mesozoic strata are generally not faulted but flexures are more common. Fractures show a non-uniform distribution and occur in subparallel clusters. Such systems are associated with aulacogens and formed at different times.

With decreasing structure size, fractures play an increasingly more important role. The smaller the structure, the shallower are its roots. Thus, it is possible to conclude that the number of fractures increases in the upper crust. This conclusion is consistent with the converse notion that plasticity of rocks increases with depth.

Types of major structures alternate regularly in space and time (Nalivkin, 1976). Aulacogens predominated in the late Proterozoic, 1400-700 Ma ago (Fig. 2). Pericratonic embayments occurred at platform margins adjacent to the active Timan and Ural geosynclines. There was only one intraplatform depression, the Orshansk syneclise.

At the very end of the Proterozoic and during early Paleozoic (700-560 Ma) aulacogens ceased to subside and much wider and gentler syneclises formed in the location of the Mid-Russian aulacogen. This was also the time of the final stage of folding and orogeny of the Timan geosyncline. Folding occurred also within the Ural geosyncline.

In early Paleozoic (560-375 Ma) (Fig. 3) aulacogens are no longer common. The emplacement of ill-defined aulacogens took place only

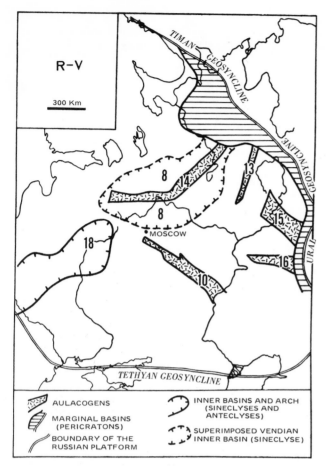

Fig. 2. Major structures of the Upper Proterozoic.
Legend: 1 = Baltic Shield, 2 = Ukrainian anteclise, 3 = Timan Ridge, 4 = Voronezh anteclise,
5 = Volga-Ural anteclise, 6 = Prekaspian syneclise, 7 = Pechora syneclise, 8 = Moscovian
syneclise, 9 = Baltic syneclise, 10 = Pachelma
aulacogen, 11 = Dnieper-Donetsk aulacogen, 12 =
Pechora-Kolvin aulacogen, 13 = Vyatka aulacogen,
14 = Mid-Russian aulacogen, 15 = Kaltasa aulacogen, 16 = Sernovodsk-Abdulino aulacogen, 17 =
Ul'yanovsk-Saratov depression, 18 = Orshansk
syneclise, 19 = Dnieper-Donetsk syneclise.

in the Ordovician and Silurian within the
Pechora syneclise and in the southeastern part
of the platform. A gentle syneclise opening
toward the west formed north of Moscow. Slow
subsidence of the adjacent geosynclines is
correlated with poor development of pericratonic
embayments.
     The formation of aulacogens and local grabens
was renewed in the early Middle and Upper Paleozoic (375-200 Ma) (Fig. 4). The vast Dnieper-
Donetsk aulacogen formed, the downwarping of the
Pechora-Kolvin aulacogen intensified and subsidence of the Vyatka aulacogen renewed but with
slight displacement. A pericratonic embayment

became clearly reflected in the topography in
the eastern part of the platform adjacent to the
actively subsiding Ural geosyncline. The formation of the vast and gentle Moscovian syneclise
continued at the site of the Mid-Russian aulacogen. At the close of the Upper Paleozoic the
subsidence of the aulacogen decreased and in
Permian time was replaced by uplift in the east
of the Dnieper-Donetsk aulacogen and within the
Pechora-Kolvin aulacogen. Concurrently, folding
and mountain building began within the Ural
geosyncline.
     During the Mesozoic and Cenozoic (375-0 Ma)
downwarping took place mainly in the southern
part of the platform adjacent to the Alpine
Tethyan geosyncline (Fig. 5). A gentle and wide
syneclise formed in position of the Dnieper-
Donetsk aulacogen. A superimposed Ul'yanovsk-
Saratov depression was formed in the middle
reaches of Volga. The close of the Cenozoic is
characterized by a final stage of folding and
orogeny in the Tethyan geosyncline and orogeny
in the Urals which was not that strong.
     Thus, during its long history the Russian

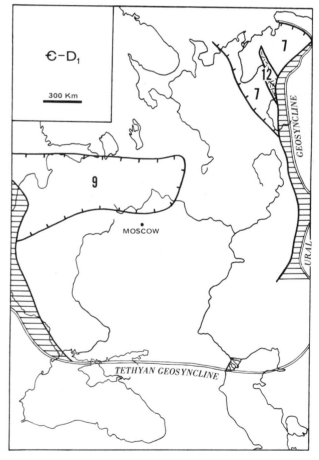

Fig. 3. Major structures of the Early Paleozoic. See Fig. 2 for Legend.

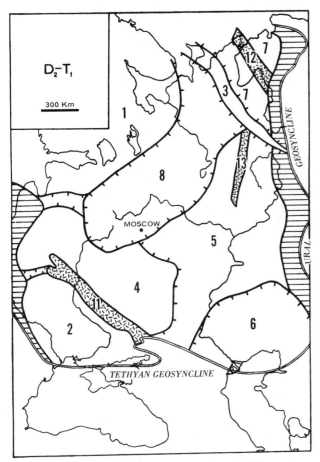

Fig. 4. Major structures of the Middle and Upper Paleozoic. See Fig. 2 for Legend.

platform alternated between periods of extensive formation of graben-aulacogens and intense downwarping of geosynclines and periods with formation of gentle synclises and anteclises as well as inverse uplifting within aulacogens and orogeny within adjacent geosynclines. The formation of pericratonic embayments took place at different places along the platform margin depending where active downwarping of geosynclines occurred at any given time.

Synchroneity of Movements

Old major marine transgressions and regressions were almost synchronous in time for the entire Russian platform (Fig. 6) (Vinogradov et al., 1974). Most of these transgressions are also known from other continents of the northern hemisphere. However, movements of platforms often were not synchronous. The Russian platform, for example, has undergone a maximum downwarping in the Middle and Upper Paleozoic. In the Jurassic and Cretaceous the subsidence was unimportant. In great contrast, the subsidence of the West Siberian platform had the

highest rate during the Jurassic and Cretaceous. The subsidence rate of the main Siberian platform to the east was very high during the Lower Paleozoic and much lower during the Middle and Upper Paleozoic, while in the Jurassic and Cretaceous was mainly a time of uplift.

However, times of high and low growth rate of tectonic structures complicating the Russian platform can be correlated. Such a correlation is also possible for marine transgressions and regressions. Rosanov (1957) stated that growth of small local uplifts was most intense during the initial stages of marine transgressions. Major structures in most cases have undergone rapid growth at the same periods of time.

Rates of Movements

Large sudden uplift and subsidence took place concurrently within a large generally subsiding area. The average rate of subsidence since the time of 1400 Ma to Recent was 7 m per Ma. The Middle and Upper Paleozoic are characterized by maximum rate of 34 m per Ma which is rather high and contrasts with the Lower Paleozoic, Mesozoic and Cenozoic with rates of 4.5 m per Ma. The rates drop to 2.6 m per Ma in the Upper Proterozoic.

The Russian platform as well as the Siberian and North American platforms belong to a northern group of ancient platforms which, as already reported, had undergone mainly subsidence. In contrast, a southern group of platforms known as the Gondwanaland group was dominantly uplifted.

The growth rate of some tectonic structures is related to their type and size. Big grabens-aulacogens subsided along fractures that are associated with some of the highest rates of subsidence among major structures. An average rate of their subsidence was about 30 m per Ma and Devonian aulacogens reached a maximum rate of 60-100 m per Ma. Inner synclises had the lowest subsidence rate, i.e., about one-fifth of that of aulacogens. The subsidence rate of pericratonic embayments was transitional or about half that of aulacogens. Thus, we have the high growth rates for structures with movements along fractures and low rates for those associated with flexure in the earth's crust and those distant from geosynclines. Geosynclines have intensified the movement at platform margins, and so the formation of pericratonic embayments situated there was rather rapid.

The decrease in the amplitude of structures associated with flexure in the earth's crust is proportional to their size. The average amplitude is about 1500 m, 500 m, 150 m and 45 m for structures within synclises; arches and depressions; swells, domes and troughs; and local structures, respectively. Because the duration of their formation is similar for all the structures, it is possible to say that the difference in growth rate is proportional to that of size.

The growth of smaller structures associated

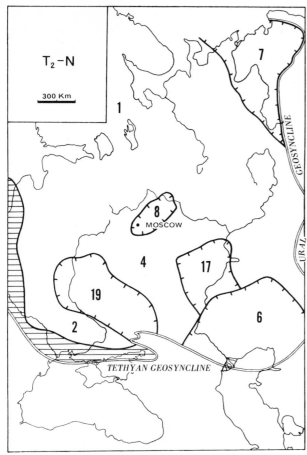

Fig. 5. Major structures of the Mesozoic and Cenozoic. See Fig. 2 for Legend.

with larger structures probably does not depend on their size. The downwarping rate of local grabens known in the east of the platform is 50-100 m per Ma, i.e., does not differ from that of major grabens-aulacogens.

During the periods of higher structural growth rates, the rate of movements was much higher than our average estimates determined for tens and hundreds of million years.

Duration of Movements

Platform structures typically grew for a very long time. The time span for a single stage of aulacogen subsidence accounted usually for 180 Ma. Devonian aulacogens (Vyatka and maybe Dnieper-Donetsk) in some cases are inherited from ancient late Proterozoic aulacogens but they became slightly displaced. The time interval between the two stages of aulacogen formation was also rather long (more than 200 Ma). The duration of subsidence of the interior syneclises was longer. It was on the average 350 Ma, with a range from 160 to 770 Ma. The pericratonic embayment life had a maximum dura-

tion; it was affected by that of adjacent geosynclines' existence and varied from 400 to 1000 Ma.

Major fractures belong to more long-lived structures. As it was stated above, non-intense movements along them continued much later when the formation of associated structures ceased.

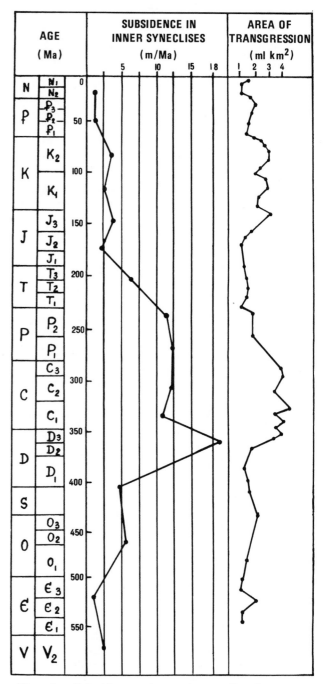

Fig. 6. Subsidence time and rate for inner syneclises and area of transgressions plotted for the Russian platform (after Nalivkin, 1976).

The duration of activity of some faults bounding old grabens-aulacogens exceeded 1400 Ma. Movements along them often changed their direction and were separated by long intervals. It was through fractures that tectonic structures affecting the sedimentary cover have apparently inherited deeper structures of the basement.

2. Dynamics of the Development of Structures of the West Siberian Platform

The basement of the West Siberian platform is much younger than that of the Russian platform. Its consolidation took place at the close of the Paleozoic. It was brought to the surface just south of the platform where rocks of different age and complex structure become quite visible. It forms the northwestern end of the extensive Ural-Mongolian fold belt that stretches eastward to the Pacific Ocean.

The old portions of the Caledonian and Baikalian basement are in places covered by slightly folded and altered Paleozoic sediments. Many geologists refer to these as the intermediate structural stage. This stage, judging from seismic data, is characterized by the continuous development in the northern part of the plate. In some depressions, it is 5-6 km thick (Fig. 7).

The platform sedimentary cover is represented by the alternation of Triassic continental strata (230 Ma) and basalt sheets. To the south they fill the grabens. The biggest graben among them (Koltogorsk-Urengoi) shows an elongated trend and lies in the middle of the platform. In the north Triassic formations build up an extensive platform cover. Despite this, some geologists assign it to the intermediate structural stage. The thickness reaches several kilometers. Its role in the history of the West Siberian platform is similar to that of Upper Proterozoic sediments of the Russian platform filling grabens-aulacogens.

Jurassic, Cretaceous, Paleogene, and Neogene sediments show continuous development (Rudkevich, 1976). Dips are of a few degrees or more often fractions of degrees. In the west they form part of the Ural geosyncline and in the east the margin of the ancient Siberian platform. The thickness of this part of the section reaches 5-6 km in the north of the platform.

Types of Tectonic Structures

The West Siberian platform is complicated by numerous structures (Fig. 8). These, like those of the Russian platform, are subdivided according to their size (Table 1). There are no pericratonic embayments on the West Siberian platform. Major grabens filled in by Triassic sediments are apparently the equivalents of aulacogens. Circular and elongate intraplatform synclises, anticlises, depressions and arches that resulted from the flexure of the earth's crust predominate. Structures related to deeper fractures are subordinate.

Fractures inside of the platform cover are rather scarce. Seismic data and a few boreholes show that in the lower horizons of the cover, fractures are rather common and disappear up the section. The distribution of elongate structures with asymmetric limbs implies the existence of several fracture systems within the sedimentary cover. One of them stretches from the southern Yamal Peninsula southeastward to the Purov graben. The second of the southwestern striking system crosses the Yenesei in the northeastern part of the platform. A third system is associated with the central Purov graben.

Synchroneity of Movements

A variation curve of marine transgression areas (Fig. 9) differs from that plotted for the Russian platform by a smaller area of transgression in the middle Early Cretaceous. In contrast, on the Russian platform, this period is marked by an increase in the area of transgression. In West Siberia this interval is characterized by a maximum rate of sedimentation which was accompanied by a mass supply of arenaceous-argillaceous material. The volume of sediments probably was so large that it filled the entire downwarping area and prevented the advance of the sea to this area. In other respects, the Jurassic-Recent transgression curves for both platforms show similar timing of changes.

Periods of higher and lower growth rates for platform structures of different size were determined using plots that compare thickness of coeval sequences between two boreholes located on the limbs and in the center of a structure. It was assumed that the sediment thickness suggests subsidence in a sedimentary basin while the thickness difference between apexes and limbs implies vertical displacement of the apex of the uplift with respect to its limb. The Neogene and in part the Paleogene are characterized by thickness differences in layers over most of these intervals. In some cases our curves overstep the zero line going to the left. In these cases the rate of sedimentation was higher on the apex than that on the limb. Two main stages of the increase in the growth rate – early-Middle Jurassic and early Cretaceous – can be recognized on curves plotted for major structures (Fig. 10). The growth rate drops during the late Mesozoic and the Cenozoic. However, some structures in the north and the east were at times subject to intense growth. A large number of reliable data obtained recently support and refine the conclusion about periodicity in growth rates (Rudkevich et al., 1976).

Despite their general similarity, plots for individual structures show distinctive patterns of their own. This is partly due to the

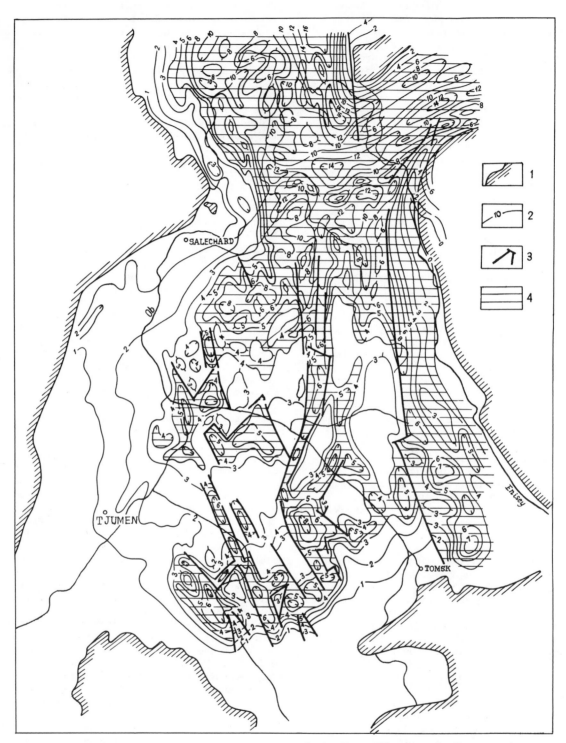

Fig. 7. Sketch map of the major structures of the West Siberian platform at the top of the basement.

Legend: 1 = Boundary of the platform, 2 = Isobaths, 3 = Fractures along the basement top and intermediate stage, 4 = Zones of intermediate stage distribution.

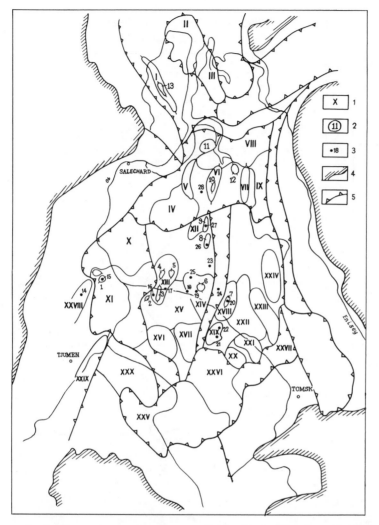

Fig. 8. Outlines of platform cover structures of the West Siberian platform. Numbers show structures for which thickness difference in the centre and on a limb is obtained. Cf. Figs. 10, 11, 12.

Major Structures. Legend: 1 = Nurmin megaswell, II = Sredne Yamal arch, III = Gidan arch, IV = Tanlov depression, VI = Urengoi megaswell, VII = Chasel megaswell, VIII = Bolschechetsk depression, IX = Srednetazov depression, X = Nadym depression, XI = Khanty-Mansi depression, XII = Purpe arch, XIII = Surgut arch, XIV = Nizhnevartov arch, XV = Yugan depression, XVI = Verchedemian arch, XVII = Kaymysov arch, XVIII = Alexandrov arch, XIX = Srednevasjugan arch, XX = Pudinsk arch, XXI = Parabel arch, XXII = Ust'-Tym depression, XIII = Pylcaramin megaswell, XXIV = Ajarmin megaswell, XXV = Omsk depression, XXVI = Njurolsk depression, XXVII = Bakchar depression, XXVIII = Shaim megaswell, XXIX = Tobol megaswell, XXX = Vagayischim arch.

Medium Uplifts. Legend: I = Endyr'dome, 2 = Salym dome, 3 = Ust'-Balyk swell, 4 = Minchim-kino dome, 5 = Fedorovka dome, 6 = Samotlor dome, 7 = Krivolutskoe swell, 8 = Vyngapur dome, 9 = Purpe, 10 = Urengoi swell, 11 = Yamburg dome, 12 = Zapolyarnoe dome, 13 = Arctic swell.

Local Uplifts. Legend: 14 = Ubinsk, 15 = Kamennaya, 16 = Pravdinsk, 17 = Surgut, 18 = Lokosovo, 19 = Megion, 20 = Okhteurie, 21 = Myl'dzhina, 22 = Severo-Vasyugansk, 23 = Variegan, 24 = Novomolodezny, 25 = Pokachivo, 26 = Vyngapur, 27 = Gobkin, 28 = Yubileinoe.

1 = Number of Major structures, 2 = Medium structures, 3 = Local structures, 4 = Boundary of the West Siberian platform, 5 = Outlines of the biggest structures.

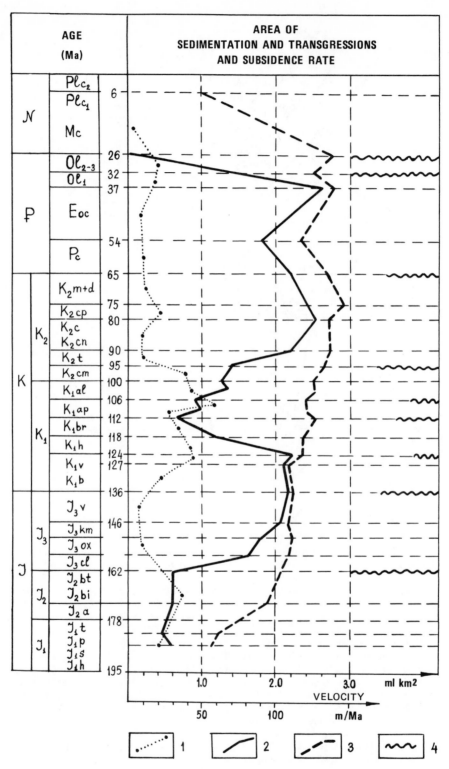

Fig. 9. Areas of transgressions and subsidence rate for the West Siberian platform.
Legend: 1 = Subsidence rate, 2 = Area of marine transgressions, 3 = Area of sedimentary basin, 4 = Breaks in sedimentation.

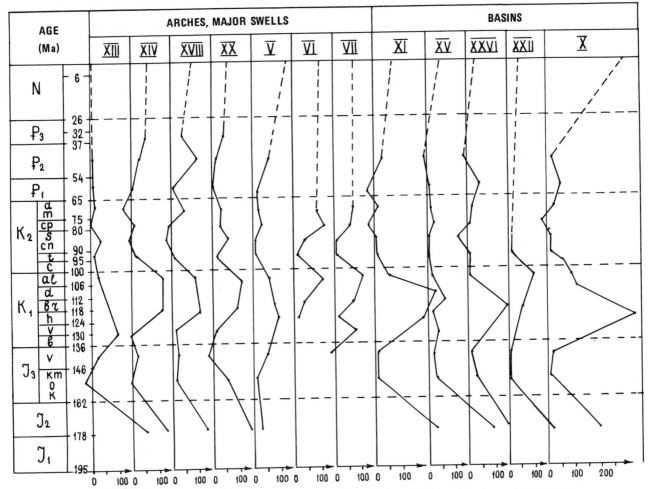

Fig. 10. Thickness difference in the centre and on limbs of major structures. See Fig. 8 for location.

irregular nature of the sedimentation process.

The position and outlines of major structures persisted throughout the entire Jurassic and Cenozoic; thickness changes imply sometimes short-term displacements of arch apexes. This may be explained by sedimentation processes.

Medium and major structures have undergone an increase of growth rates during the same periods of time, i.e., in the Upper and Middle Jurassic (Fig. 11). In the late Cretaceous and Cenozoic, the growth rate drops; however, in the north, structural growth was at that time more intensive than in the south, and increased by 130-150 m.

Plots for local structures are not so easy to correlate as those of major and medium structures (Fig. 12). However, the periods of increase and decrease in the growth rates remain almost the same, and northern structures are also characterized by higher growth rates in the Cenozoic.

Thus, periods of increase and decrease of the growth rates were in general the same for struc-

tures of all sizes. The periods of high rates of growth coincide with marine regressions of the Early and Middle Jurassic as well as those at the close of the Early Cretaceous. The situation is not that obvious for the regression which began at the close of the Tertiary. The northern part of the plate was most intensely uplifted in the Neogene as well as on the adjacent Taimyr and northern Siberian platform. The growth rate of structures decreased during transgressions.

Regional and zonal changes affect the platform cover (Fig. 9). Many of them are related to the changes of the curves of sedimentation rate versus the curves for marine transgression areas. They increase in number during regressions. Their position in the section is comparable to that of the Russian platform.

Rate of Movements

The above discussed high and low subsidence rates of the entire platform caused breaks in

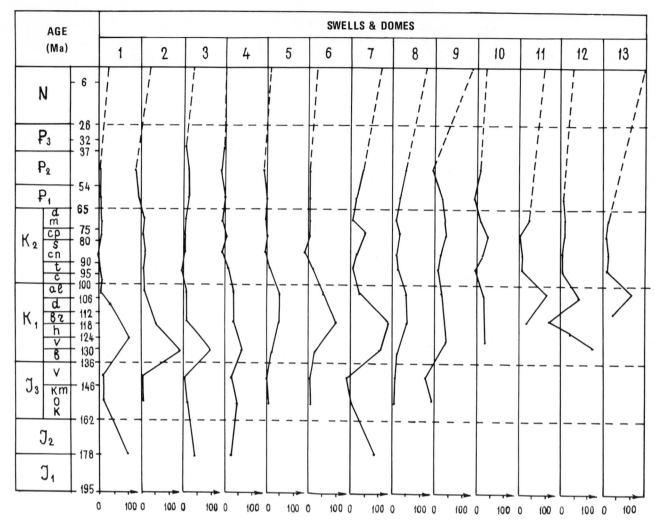

Fig. 11. Thickness difference in the centre and on limbs of medium structures. See Fig. 8 for location.

sedimentation that occurred during an overall accelerated subsidence of 20-25 m per Ma from the Jurassic to the Neogene which is three or four times the average subsidence rate of the Russian platform.

Duration of Movements

Many structures that complicate the platform cover of the West Siberian platform inherit the strike and partly the shape of structures recognized in Triassic formations. Also some major structures match the outlines of magnetic and gravity anomalies and enable us to consider these anomalies as related to the inner structure of the basement. All these suggest movements that were inherited from a period exceeding 200 Ma. The growth of structures still continues today.

3. Deep-seated Structure Based on Geological and Geophysical Data

The formation of major platform structures of 100-200 km in diameter is accounted for by processes operative at great depth in the earth's interior. Their loci are at a depth of several tens of kilometers and may be still deeper (Klushin et al., 1970). This is also evident from the nature of intracrustal seismic interfaces and especially from the shape of the base of the earth's crust. Thus, the knowledge of the earth's crust and upper mantle structure is important if we are to understand the nature of sedimentary basins and uplifts that separate them.

The profiles (Figs. 13, 14, 15) were made in different years and by various institutions. They were compiled in a single profile: the Russian platform is underlain by the early

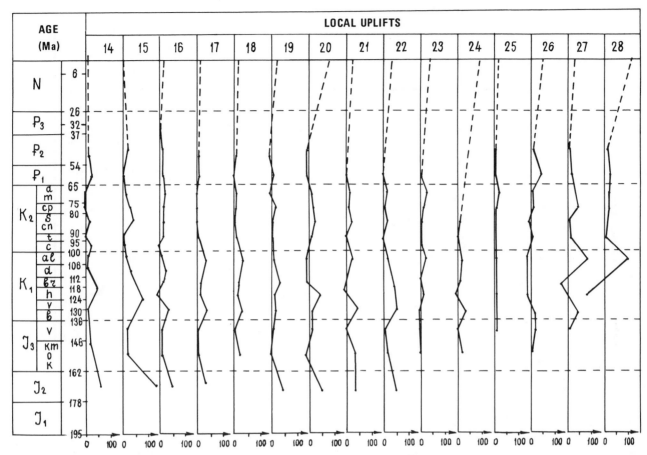

Fig. 12. Thickness difference in the centre and on limbs of local structures. See Fig. 8 for location.

Proterozoic basement (1700–1900 Ma) which includes cores of Archean age (2500–2800 Ma); the Urals resulted from the Hercynian folding (230 Ma); the West Siberian platform underlain is by a Caledonian (400 Ma), by an early as well as a late Hercynian (230 Ma) basement and by a core of Baikalides (570–700 Ma). The Yenesei Ridge belongs to the Baikalide folds (570–700 Ma) and the western margin of the Siberian platform has a basement older than 1700 Ma.

The platform basement, especially that of the Precambrian age such as the Russian platform, is built up by intensely folded metamorphosed rocks. Gentle primary layers are not likely to exist inside of the basement. Nevertheless, seismic profiles show gentle reflectors traceable for more than 300 km for both the "granite" and "basalt" layers. Similar reflectors are also known for the ancient Siberian platform. Their nature is not quite understood. They may suggest a varying degree of compaction of the crustal matter by the heavy overlying rocks.

The profiles shown, as well as other data, suggest that the thickness of the earth's crust is from 37 to 53 km (on the average, 46 km); 40–48 km (on the average, 43 km); 33–45 km (on the average, 38 km) on the Russian platform, the Urals, West Siberian platform, Yenesei Ridge and Siberian platform, respectively. Thus, the earth's crust thickness for the West Siberian platform is a little less, probably due to its deep subsidence.

An average thickness of the crust for all the platforms on the earth according to Ronov and Yaroshevsky (1976) is 43.2 km.

The inverse relation between thickness of the sedimentary platform cover and that of the earth's crust is established for the Russian platform. As a rule, the more the thickness of a sedimentary cover, the less is that of the underlying crystalline crust. Thus, crustal basement thickness averages 38 km, 41 km and 46 km when thickness of a sedimentary cover is >5 km, 1–5 km, and 1 km, respectively. This trend is not so obvious on the West Siberian profile (Krylov et al., 1974; Fig. 15).

In the central and southern parts of the West Siberian platform where the thickness of the Mesozoic and Cenozoic cover does not exceed

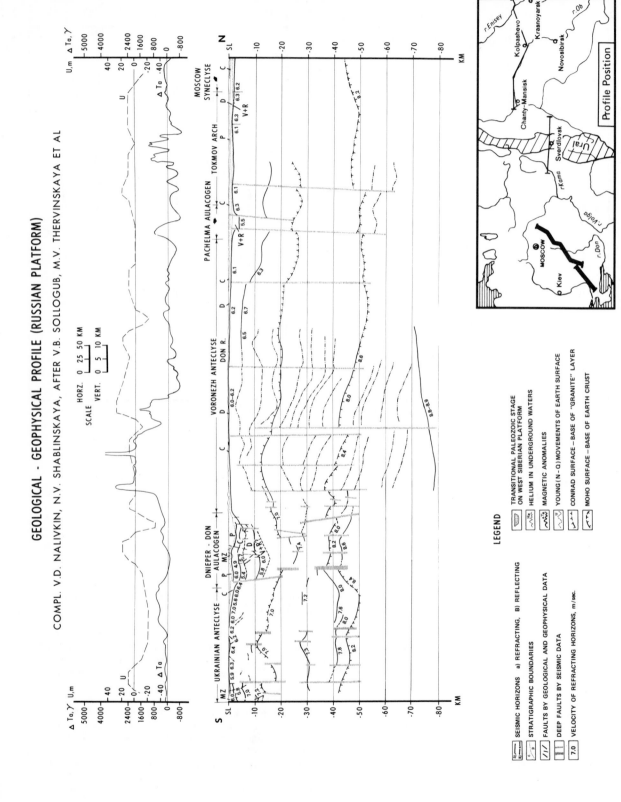

Fig. 13. Geological-geophysical profile across the Russian platform.

GEOLOGICAL - GEOPHYSICAL PROFILE (URAL)

COMPL. Yu.P. BULASHEVITCH, V.N. BASHORIN, V.S. DRUZININ, V.M. RYBALKA,
AFTER V.S. DRUZININ ET AL (1968—1969)

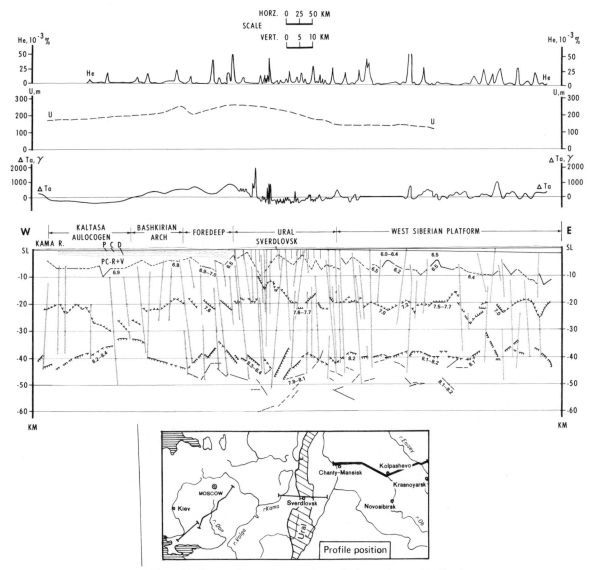

Fig. 14. Geological-geophysical profile across the Urals.
See Fig. 13 for Legend.

3.5 km, the crustal thickness is smaller under the Paleozoic depressions. In the north the crustal thickness is affected by the thickness of the Upper Paleozoic and Mesozoic-Cenozoic sediments. In areas where the sediment thickness exceeds 8 km, the crustal thickness goes down to 33 km. A displacement and even discrepancy between sedimentary cover structures and those of uplifts and depressions at the crust base are inferred for the Siberian platform. The displacement of structural frameworks is identified also on the Russian platform just

north of the Dnieper-Donetsk aulacogen, east of the Kaltasa aulacogen and in other places. To be more correct, one should speak about persistent trends and not about strict fit.

Isostasy, i.e., hydrostatic balance of lithosphere blocks, is one of the possible reasons. The base of the earth's crust (M-discontinuity) is entirely inside of the lithosphere and in a way responds to the movements of its blocks. Within blocks where sedimentary rocks account for a considerable part of the earth's crust, its basement thickness decreases because a

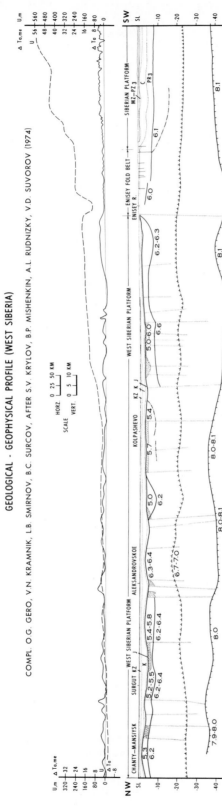

GEOLOGICAL - GEOPHYSICAL PROFILE (WEST SIBERIA)

COMPL. O.G. GERO, V.N. KRAMNIK, L.B. SMIRNOV, B.C. SURCOV, AFTER S.V. KRYLOV, B.P. MISHENKIN, A.L. RUDNIZKY, V.D. SUVOROV (1974)

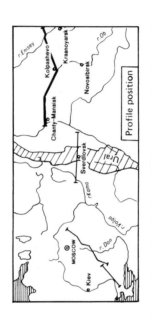

Fig. 15. Geological-geophysical profile across the West Siberian platform.
See Fig. 13 for Legend.

basalt layer appears to be converted to the mantle compensating for the mass deficiency due to sedimentary rocks. This phenomenon became more evident in the oceans where a thick water body is compensated as indicated by the nearness of the M-discontinuity to the geoid surface.

The interface of the "granite" and "basalt" layers remains uncertain. Maybe it is a transition zone. On the profiles, it is tentatively indicated.

It is interesting to note that the interface of the "granite" layer corresponding to that of the Russian platform basement on the basis of seismic and magnetic data is traced under folds and nappes on the western slope of the Urals and eastward to the dividing ridge coinciding with a major fault separating miogeosynclinal and eugeosynclinal zones. The folding is developed there above the unfaulted basement.

There is no ubiquitous relation between late movements (N-Q) and the thickness of the earth's crust. We can speak about a great variety of relationships. Zones of late (N-Q) uplifts coincide with zones of larger crustal thickness and positive structures only in the Urals, Yenesei Ridge and in the northern part of the Siberian platform where movements become more evident.

The Urals are characterized by positive isostatic anomalies. Hence, modern uplifts are anti-isostatic and result from active tectonism. The uplift probably was affected by gravity. This is shown by geophysical data and measurements of rock stress in pits which show the increase in absolute value of latitudinal compression and shear stress (Aleinikov et al., 1967).

The inverse relation is observed for the Dnieper-Donetsk aulacogen. Young uplifts coincide there with a zone of small crustal thickness (Fig. 13). Such a relation may be due to the fact that this graben at the close of the Permian and during the Mesozoic was subject to inverse uplift; however, that uplift was less intensive than the preceding subsidence and so the graben structure of the earth's crust has not been changed greatly. The uplifts probably occurred during times of compression.

The profiles (Figs. 13, 14, 15) show many near-vertical faults. They are recognized on the basis of the lack of correlation between seismic boundaries and vertical displacements of such boundaries. The faults go to great depths as established by the seismic evidence. They often coincide with distinct linear magnetic and gravity anomalies. Such anomalies are believed to be caused by the development of ultrabasic and basic magmatic rocks intruded along deep fractures. Many long-lived fractures are visible on areal photographs and images.

It is noteworthy that a number of fractures are marked by displacement of all the refractors which implies that they can be traced to great depths.

The permeability associated with active and renewed movements along fractures in the Urals and adjacent areas also may be recognized on the basis of high content of helium dissolved in underground waters. Water samples were taken mainly from boreholes at a depth of 60-120 m every two to three kilometers within a belt 10 km wide following deep seismic profile. Helium background concentration amounts to $0.5-2.0 \times 10^{-3}$ volume percent. All high helium measurements are unambiguously associated with fractures on seismic profiles (Fig. 14). However, there are a number of fractures recognized on the seismic surveys that have no helium anomalies. Some fractures are apparently healed and become impermeable for helium (Bulashevich et al., 1973).

Finally, a large number of ultrabasic intrusions related to fractures also argues for their deep-seated structure that can be established in the field.

A majority of fractures parallel the structural trends of the platform basement, and thus suggest their old age. Movements continue to take place along many of them. Fractures formed on platforms heal with difficulty. However, intersecting younger superimposed fractures also can be observed. Let us first mention rift zones of the West Siberian platform that cut Paleozoic folded structures almost at right angles. The Kiltogor graben-rift and its apophyses is an example of this type.

Faults have non-uniform areal distribution and form clusters. This is quite evident on the profile crossing the Urals (Fig. 14). The highest density of faults occurs in the Ural eugeosynclinal zones. It is also within this zone of fault clusters that anomalously high helium contents in underground waters are measured. There the number of faults permeable for helium is four per 50 km of a profile eastward and westward it is one per 50 km.

Faults and other linear tectonic features of platforms belong mainly to two systems of trends, i.e., orthogonal and diagonal.

Conclusions

1. On the Russian, West Siberian and some other platforms, sedimentation of the cover rocks has begun with the formation of major grabens-aulacogens. Structures of this type are apparently most common at initial stages of platform development. Therefore, we do agree with Khain (1973) who proposed to distinguish an initial aulacogen stage of platform development.

2. There is a temporal connection between the aulacogen formation and geosyncline subsidence as well as the cessation of aulacogen downwarping and partly uplift and orogeny and folding inside of geosynclines. The aulacogen and geosyncline downwarping is associated with epochs of major transgressions while movements

of inversion type of aulacogens and orogeny are associated with epochs of regressions. As Milanovsky (unpublished) stated, there is possibly a relation between the periods of overall compression and expansion of the earth.

3. Most platform cover structures of different size both on ancient and young platforms suggest an increase in growth rate during regressions and decrease in growth rate during transgressions. All this implies the existence of uniform periodical processes operative over vast areas of the globe.

4. Average growth rates of platform structures are very low, i.e., 0.15-5 m per Ma. On the ancient Russian platform, they were smaller than those on the young West Siberian platform. Local structures show very great differences in growth rates. Maximum subsidence rates of the entire Russian platform were on the whole lower than those of the West Siberian platform and did not exceed several tens of meters per Ma.

However, short-term movements were rather rapid. The rate of uplift and subsidence of some regions determined using repeated leveling reaches 1-2 cm per year of 10,000-20,000 m per Ma. These movements are very short-term and undoubtedly have quite different explanations.

5. The duration of structural growth can be rather long. Both on the young West Siberian and ancient Russian platforms, structures continue to grow and many of them are today reflected in the topography. The formation of more long-term pericratonic embayments continued for about 1 Ga. The same is true for the Baltic and Ukrainian shields. The duration of their life accounts for 18 percent of that of the earth. It suggests that deep-seated loci of tectonic movements are persistant both in time and with respect to platforms. Major fractures have even longer duration of life. They account for 25 percent of the earth's age.

However, the duration of growth and development of structures is limited. There is no doubt as to the emplacement of new major structures (Lower Paleozoic Baltic and Mesozoic Ul'yanovsk-Saratov syneclises) and cessation of growth of ancient structures (Pachelma and Sernovodsk-Abduline aulacogens). The most essential reconstruction occurs during the overall major regressions with periodicity of 180 Ma.

6. Fractures on both platforms discussed are most widespread near the top of the basement. They decrease in number in the upper parts of a sedimentary cover and apparently in the lower crust.

7. Deep seismic soundings clearly show that the loci of tectonic movements responsible for the formation of major and biggest structures both on the ancient and young platforms lie below the earth's crust, i.e., the M-discontinuity. Their evolution should be concurrent with those of lithosphere plates.

8. The presence of steep fractures is another argument against permanent and considerable horizontal displacements in the earth's crust. At the same time nappes with a displacement of tens of kilometers were also proven to exist in the upper crust and in the Urals as well. The duration of nappe movement is rather short while movements along vertical faults continue for a very long time and are sometimes renewed when the formation of nappe ceases. This may account for their existence.

However, the presence of the fractures does not obviate the possibility for plate movement if it occurs in deeper layers (Shablinskaya, 1977).

Non-intensive horizontal movements probably take place within plates as well. The argument for this is the formation of tension structures in rift-grabens and magnetic data showing the presence of transform faults within continental plates.

References

Aleinikov, A. L., O. V. Bellavin, and I. F. Tavrin, K voprosu o tectonicheskom razvitii Urala. V. knige "Stroenie zemnoy kory Urlla," Sverdlovsk (in Russian), 1967.

Belyaevsky, N. A., and A. A. Borisov, Structura i moshchnost' zemnoi kory SSSR, in book Structura fundamenta platformennykh oblastei SSSR, Nauka, Leningrad, 381 pp. (in Russian), 1974.

Bronguleev, V. V., i dr., Karta reljefa raznovozrastnogo fundamenta Vostochno-Evropeiskoi platformy, 1:2,500,000 (6 layers, in Russian), 1975.

Bulashevitch, Yu.P, V. N. Bashorin, V. S. Druzhinin, V. M. Rybalka, Heliy v podzemnych vodach na Sverdlovskom profile glubinnych seismicheskikh zondirovaniy, Doklady AN SSSR, ser. matematich. fisk., t. 208, N 4, 1973.

Druzhinin, V. S., L. N. Kazachichina, V. T. Politov, V. M. Rybalka, N. I. Khalevin, and L. N. Thudakova, Glubinnoe stroenie zemnoy kory Urala i prilegajuschikh k nemu oblastey po Sverdlovskomu subschirotnomu peresecheniyu, in book Magmaticheskie formatsiik metamorfizm, metallogeniya Urala, t. 1, Sverdlovsk, 1969.

Druzhinin, V. S., V. M. Rybalka, and N. I. Khalevin, Rezultaty glubinnykh seismicheskykh zondirovaniy na Sverdlovskom peresechenii i perspectivy dalneishikh issledovaniy Urala. Sb. "Glubinnoe stroenie Urala," Nauka, M., 69-80 (in Russian), 1968.

Khain, V. E., Obshchaya geotektonika, Nedra, M., 509 pp. (in Russian), 1973.

Khalevin, N. I., Seismologiya vzryvov na Urale, Nauka, M., 133 (in Russian).

Khatyanov, F. I., O tectonicheskoi prirode pogrebennykh devonskikh micrograbenov i perspectivakh poiskov neftenosnykh structur na yugo-vostoke Russkoi platformy., Geol. Nefti i Gaza, 7, 41-46 (in Russian), 1971.

Klushin, I. G., V. D. Nalivkin, and N. V. Shablinskaya, Ozenka polozheniya kornei platformennykh struktur, in book Problemy stroeniya zemnoi kory i verchnei mantii, Nauka, M., 131-135 (in Russian), 1970.

Krylov, S. V., Mischen'kin B.P. i dr., Charakteristika Aapadno-Sibirskogo regiona i dannye glubbinogo seismicheskogo zondirovaniya, in book Stroerie zemnoi dory v Zapadnoi Sibiri, Nauka, Novosibirsk, 7-15 (in Russian), 1974.

Nalivkin, V. D., Dynamics of the development of the Russian platform structures, Tectonophysics, 36, 247-262, 1976.

Ronov, A. B., and A. A. Yaroshevsky, Novaya model chimicheskogo stroenia zemnoi kory, Geochimia, , 12, 1763-1795 (in Russian), 1976.

Rozanov, L. N., Istoriya formirovaniya tektonicheskih struktur Bashkirii i prilegayaschchih oblastei., Gostoptehizdat, M. 207 (in Russian), 1957.

Rudkevich, M.Ya., The history and the dynamics of the development of the West Siberian platform, Tectonophysics, 36, 275-287, 1976.

Rudkevich, M.Ya, Yu.M. Glukhoedov, and E. M. Maksimov, Tectonicheskoe razvitie i neftegeologicheskoe rayonirovanie Zapadno-Sibirskoi provintsii., Sredneuralskoe knizhnoe izdatel'stvo, Sverdlovsk, 170 (in Russian), 1976.

Shablinskaya, N. V., Novye dannye o globalnoi setke razlomov na planete, Doklady AN SSSR, t. 237, N 5, 1159-1162 (in Russian), 1977.

Vinogradov, S. P. ed.-in-chief, Paleogeographiya, SSSR, (4 tome), Nedra, Moscow (in Russian), 1974.

Vol'vovsky, I. S., and B. S. Vol'vovsky, Razrezy territorii SSSR po dannym glubinnogo seismicheskogo zondirovaniya, Sovetskoe radio, M., 260 (in Russian), 1975.

BASINS OF THE AUSTRALIAN CRATON AND MARGIN

J. J. Veevers

School of Earth Sciences, Macquarie University
North Ryde, NSW 2113, Australia

Abstract. The history of modern sedimentary basins of Australia is related to events that take place at present or past plate boundaries: the foredeep between the Australian mainland and New Guinea is bounded by uplift of the New Guinea Highlands along a convergent collisional plate boundary: the internal basins of the Lake Eyre and Murray River regions derive their sediment from the eastern and southeastern margins that were uplifted during plate divergence in the Late Mesozoic and Early Cenozoic; and the peripheral basins of the western, southern and eastern margins lie across collapsed arches that were split by plate divergence in the Mesozoic and Cenozoic. Similar patterns in the ancient basins of Australia can be related to similar causes. A notable departure from global sea level is found in Australia during the middle Cretaceous (~100 Ma ago), when an epeiric sea covered much of Australia at the same time as rapid convergence took place along the eastern margin; the sea retreated in the Late Cretaceous, at a time of decayed local convergence, but at the same time as global sea level reached a maximum. Both subsidence and the succeeding uplift of the interior are thus due to convergence and its decay along the adjacent margin. Other similar events in older sedimentary basins are traceable to effects at plate boundaries.

Introduction

The pattern of modern sedimentary basins on the Australian craton and its margins (Fig. 1) is a useful guide to past patterns of basin development in this region. The northern margin is bounded by a mountain range, up to 5 km high (New Guinea and the Outer Banda Arc, principally Timor), which supplies sediment to an adjacent foredeep, and, together with the northern part of mainland Australia (Doutch, 1976), contributes sediment to a broad foreland basin (Timor and Arafura Seas and the Gulf of Carpentaria); in the eastern part of this basin, terrigenous sediment dominates, and in the west it is diluted to a greater or lesser degree by autoch-

thonous carbonate sediment (Heckel, 1972, p. 269). To the south, two internal basins accumulate sediment shed from a low (<2.3 km high, generally <1 km high) range of mountains along the eastern seaboard, called the Great Dividing Range. And, on the west, south, and east, the margins accumulate prograding sediment, mainly carbonate.

The northern foredeep and foreland basin, which owe their existence to the convergence of the Australian Plate and the Eurasian and Pacific Plates (Audley-Charles et al., 1977, Fig. 2), dominate the region, both in their extent and volume of sediment. The other basins cover large areas but accumulate sediment very slowly; the sediment of the internal basins is entirely terrigenous, and that of the peripheral basins, which owe their existence to plate divergence in the Mesozoic and Paleogene, is almost entirely biogenous.

The foreland basin and the peripheral basins are obviously relatable to events that occur at plate boundaries; the origin of the internal basins is less obvious but probably they too are relatable to events at plate boundaries (Veevers and Rundle, 1979): arching of the eastern and southeastern margins developed in the Paleozoic Tasman Fold Belt before and during plate divergence during the late Mesozoic and early Cenozoic. They represent extra-arch basins (Veevers and Cotterill, 1978).

The variety yet simplicity of basin development in Australia, present and past, make it a good area for regional studies of basins and of the plate activity that generates them. Prerequisite to an analysis of the development of Australian basins is an understanding of the unity of the Australian craton and its setting, during most of the period under review, in Gondwanaland.

Unity of the Australian Craton and its Setting in Gondwanaland

Three important conclusions concerning the ancient configuration of Australia can be drawn from paleomagnetism (Veevers and McElhinny,

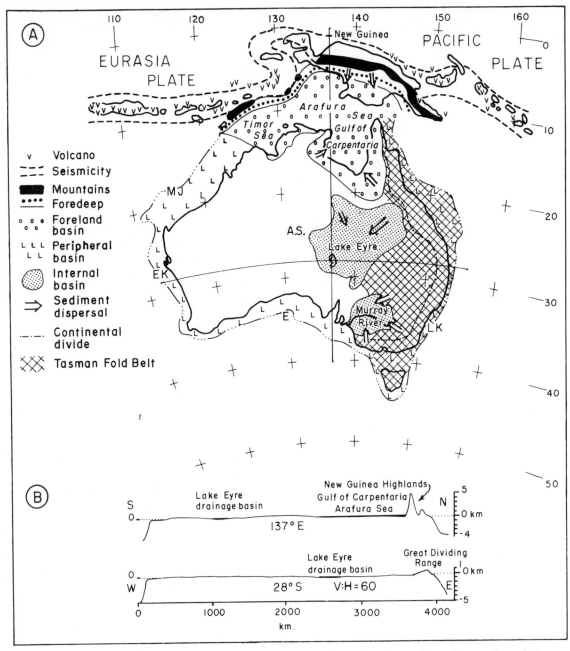

Fig. 1. (A) Continental block of Australia within the Australian Plate, bounded to the north by the Eurasian and Pacific Plates. The block is rimmed by mountain ranges in the north and east which shed sediment back toward the continental interior. The other main basins lie along the western, southern, and eastern periphery. MJ = Middle Jurassic; EK = Early Cretaceous; E = Eocene; LK = Late Cretaceous; all referring to age on inception of margin, AS = Alice Springs.

(B) Profiles of surface elevation along longitude 137°E and along latitude 28°S, showing peripheral mountain ranges on the north and east, and the broad internal depressions of the Arafura Sea-Gulf of Carpentaria and Lake Eyre drainage basin.

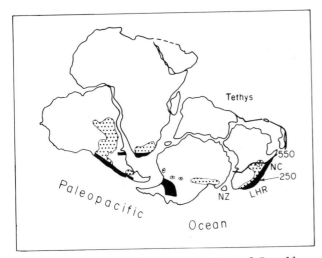

Fig. 2. Gondwanaland reconstruction of Powell et al. (in press) showing those Permo-Triassic (250-200 Ma) Gondwanan basins (dots) yoked to Late Paleozoic-Early Mesozoic mountains (solid) along the Paleopacific margin. The accretion of Australia from its eastern margin at 550 Ma ago to the present shoreline, reached by about 250 Ma ago, is shown. NZ = New Zealand; LHR = Lord Howe Rise; NC = New Caledonia.

1976): (1) the various Precambrian blocks that comprise the Australian craton have maintained the same relative position from 1800 to 750 Ma ago; (2) the Australian platform, comprising these blocks together with their sedimentary cover (this excludes the Tasman Fold Belt that now makes up the eastern third of the continent) has been united from 750 to 450 Ma ago; and (3) Gondwanaland, comprising South America, Africa, India, Antarctica, and Australia, existed at least 750 Ma ago until its disintegration about 150 Ma ago. An account of the Australian basins is thus explicable only in terms of its setting in Gondwanaland.

A reconstruction of Gondwanaland at about 150 Ma ago that takes into account ocean floor studies of the past few years (Powell et al., in press) is shown in Figure 2. What are now the western and southern margins adjoined neighboring parts of Gondwanaland, the northwestern and northern margins (excluding the later accretion of the northern part of New Guinea) faced Tethys, and the eastern margin, including the Lord Howe Rise and New Zealand, which accreted from its end of Precambrian position to its present shoreline by about 250 Ma ago, faced the Paleopacific Ocean. It is in this configuration that all but the last two tectonic sketches of Figure 3 should be viewed.

Since the presumed inception of the Pacific margin of Australia at the beginning of the Phanerozoic (570 Ma ago), the eastern (and later the northern) accreting margin has lain a greater or shorter distance from a plate bound-

ary; most of the time the boundary has separated converging plates; at other times, notably during its presumed inception and during the period 80 to 60 Ma ago, when the Tasman Sea was generated by seafloor spreading, the boundary separated diverging plates. Until the breakup of Gondwanaland the only other plate boundary in the region lay off northwest Australia 550 to 450 Ma ago. Thereafter, initially the western and central part of Australia and subsequently, with eastward accretion, the eastern part occupied the interior of a plate. For a short time in the late Mesozoic and early Cenozoic, during the inception of the western and southern margins, these regions lay along plate boundaries. With continued plate accretion, they now lie again within the interior of a plate, and, with the notable exception of its northern margin in Timor and New Guinea, Australia today lies wholly within a plate interior. Consequently, its basins, modern and ancient, reflect the vertical movements of a plate interior.

Review of Basin Development

Figure 3 summarizes the main lines of development of Australian basins.

The first phase distinguished is that during which the Adelaidean succession of monotonously shallow-water sediments was deposited over the central part of the present continent, probably from at least 800 Ma ago (Cooper, 1975) to 600 Ma ago. The craton had become a rigid unit and subsided uniformly to be covered by the Adelaidean System. According to Brown et al. (1968, pp. 39, 40), "the Adelaidean is apparently entirely devoid of greywacke-type sediments... deposition took place under shallow marine or terrestrial conditions." Older Precambrian blocks lay eastward. The position at this time of the eastern margin of Australia is unknown except that it lay an unknown distance east of the line that joins the easternmost known Precambrian rocks (Fig. 3A).

The quiescence of this phase was terminated abruptly in the interval 600 to 535 Ma ago (Fig. 3B) by deformation, notably in central Australia, by an outburst of magmatic activity on the platform (plateau basalts) and to the east, and by a wide marine transgression over the platform (among others the Georgina and Arafura Basins) and the deposition of thick greywacke-suite sediments (Kanmantoo Group) near Adelaide. All these events are taken as indicating the inception of the eastern (Pacific) and northwestern (Tethyan) margins of Australia by plate divergence. The timing of these events at the boundary between the Cryptozoic and Phanerozoic Eons is precise and is based on the Vendian Ediacara fauna, which preceded these events, and the Early Cambrian fauna, which postdated them.

Failed arms of triple junctions penetrated deep into the interior from both the Tethyan

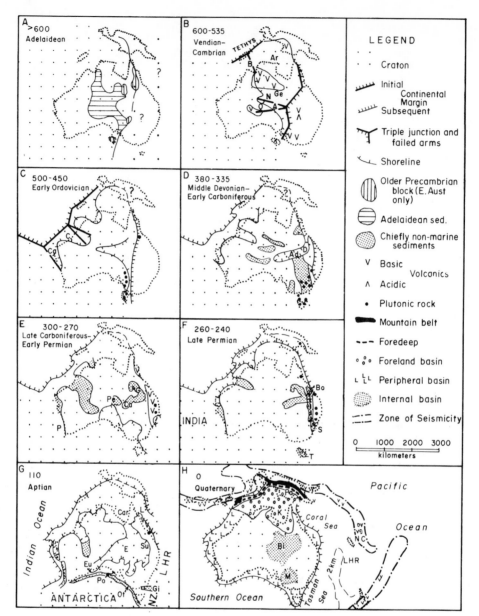

Fig. 3. (A) to (G). Paleogeographical and tectonic sketches of Australia (A to D) modified from Veevers, 1976; (E - G) from Veevers and Evans, 1975, and (H), present geographical setting. On (A) the line from New Guinea to Tasmania indicates the possible eastern extent of the Australian craton. In (A - C), the position of Tasmania follows the scheme of Harrington et al. (1973). Selected basins labeled (first appearance only) as follows: A = Amadeus; Ad = Adavale; Ar = Arafura; B = Bonaparte Gulf; Bi = Birdsville; Bo = Bowen; C = Canning; Ca = Carnarvon; Car = Carpentaria; Co = Cooper; D = Drummond; E = Eromanga; Eu = Eucla; G = Galilee; Ge = Georgina; Gi = Gippsland; K = Karumba; L = Laura; M = Murray; N = Ngalia; O = Officer; Ot = Otway; P = Perth; Pa = Papuan; Pe = Pedirka; Po = Polda; S = Sidney; Su = Surat; T = Tasmania.

(Bonaparte Gulf Basin) and Pacific (Amadeus, Ngalia, Officer Basins) margins. With the extension of the Tethyan margin 500 to 450 Ma ago (Fig. 3C), further failed arms (Canning, Carnarvon Basins) penetrated deep into the craton; the sea transgressed across the Canning Basin, and remained in the same area since early Middle Cambrian (540 Ma ago) in the Bonaparte

Gulf, Georgina, and Amadeus Basins, but altogether the sea covered a smaller area than that of the early Middle Cambrian. Deformation and plutonic intrusion of the southeast margin were probably the result of a reversal of plate motion, in which the Pacific Plate converged with Australia, and incidentally, with neighboring Antarctica, the Transantarctic Mountains of which are dotted with plutons of this age.

With continual accretion of arcs to eastern Australia (Packham, 1973; Scheibner, 1973), the eastern margin almost reached the present coastline by the Early Carboniferous (335 Ma ago) (Fig. 3D). With widespread uplift of central Australia caused by closure of the failed arms of the Amadeus and Officer Basins by crustal thrusting, the sea retreated to a few pericontinental indentations.

With the Late Carboniferous and Early Permian (300-270 Ma ago) (Fig. 3E), following a rapid change from middle to high latitudes, Australia underwent a marked change with the inception of the non-marine Gondwana-type basins (Perth, Pedirka, Cooper, Galilee), the rejuvenation of others (Carnarvon, Canning, Bonaparte Gulf), and a marine transgression across part of southern Australia. The non-marine Gondwana-type basins contain a characteristic facies of fluviodeltaic sediments with glacial and peri-glacial features and with abundant coal. The inception and rejuvenation of basins on the west anticipated the subsequent formation of rift valleys that marked the locus of plate divergence in the late Mesozoic. Widespread volcanism and plutonism on the east followed the well-established pattern of plate interaction with the Pacific.

By the end of the Permian (Fig. 3F), the basins had contracted, and in the east a long foredeep or back-arc basin (Sydney and Bowen Basins) (GSA, 1971) accumulated thick sediments shed from a highland fold belt along the (present) eastern margin, part of a chain of linear foredeeps and broader foreland basins along the Paleopacific margin of Gondwanaland (Fig. 2).

What was to become the western margin was a system of rifted arches in the early Mesozoic, and they culminated in plate divergence along the northwest margin in the Late Jurassic (160 Ma ago) and along the southwest margin in the Early Cretaceous (125 Ma ago) (Exon and Willcox, 1978; Veevers and Cotterill, 1978). Also in the Late Jurassic (160 Ma ago), what was to become the southern margin (Boeuf and Doust, 1975; Deighton et al., 1976) was blocked out by a system of rifted arches accompanied in places by profuse outpourings of basalt; by this time, much of the central and eastern part of Australia had become a broad area of mainly fluvial deposition, called the Great Artesian Basin, and comprising the Carpentaria, Surat and Eromanga Basins. By the middle of the Cretaceous Period (Aptian, 110 Ma ago), a broad epeiric sea covered this area and extended farther southwestward to the Eucla Basin, and the sea entered

the Canning Basin and the depressions (rim basin) behind the dismembered marginal arches; along what was to become the southern margin, non-marine sediment continued to accumulate in graben (Polda Trough, Otway and Gippsland Basins) set within long arches. During the short duration of the epeiric sea, calcalkaline volcanics and plutonics were emplaced along an arc along the northeastern margin and contributed sediment to the adjacent epeiric basins (Exon and Senior, 1976). Shortly thereafter, during the Cenomanian (100 to 92 Ma ago), the epeiric sea left the continent, and only the northern, western and southern margins (excluding the Gippsland Basin) remained marine.

Modern Australia (Fig. 3H) took its present shape with the dispersal of the Lord Howe Rise and New Zealand by plate divergence from 80 to 60 Ma ago, of Antarctica by plate divergence from 55 Ma ago, and of the Papuan Peninsula by the opening of the Coral Sea 55 Ma ago and the obduction of the Papuan Ultramafic Belt; and New Guinea collided with the Pacific Plate 30 Ma ago (Brown et al., 1975; Page, 1976; Thompson, 1972), and Timor with the Eurasian Plate about 5-10 Ma ago (Carter et al., 1976; Chamalaun and Grady, 1978; Powell and Johnson, in press).

Plate collision in the north led to marginal uplift behind a linear foredeep and broad foreland basin; the arching that accompanied plate divergence on the eastern margin persisted, probably due to volcanic activity related to migration of Australia northwards over a mantle magma source (Wellman and McDougall, 1974a,b; Sutherland et al., 1978), and shed sediment into internal extra-arch basins; elsewhere the margins slowly subsided and accumulated thin biogenous, mainly calcareous, sediment (Quilty, 1977, in press).

Vertical Motions of the Plate Interior

Much of the foregoing description has necessarily concentrated on the large-scale horizontal motions of plates. In seeking explanations of the vertical motions of the plate interior, as indicated by the development of cratonic basins, we find that the vertical motions are correlatable with horizontal motions. Before pointing to these correlations, we survey global trends in sea-level changes and the vertical motions of other continents.

A curve of global sea-level changes (from Vail et al., 1977) is shown in Figure 4E. It is based on two kinds of information: back to the Jurassic (0 to 200 Ma ago), the seismic stratigraphy of the margins of all the continents except Antarctica, reflecting directly the rise and fall of sea level; and from the Triassic to the Cambrian (200 to 570 Ma ago), the positions of the craton with respect to sea level, as shown by the proportion of the North American craton covered by marine deposition, with supporting data from other continents, in par-

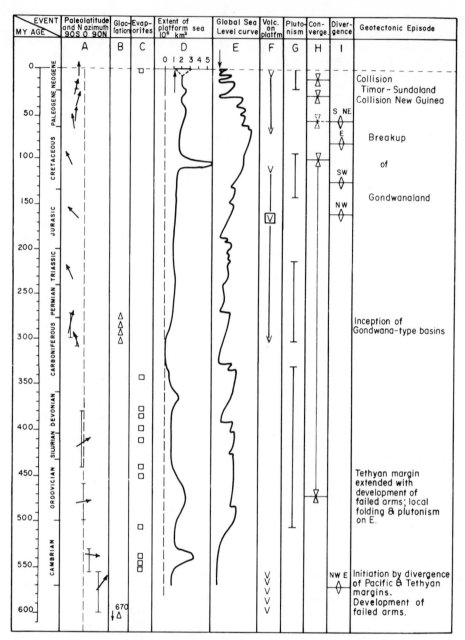

Fig. 4. Summary of selected events on the Australian platform and at the plate boundaries. Paleolatitude, referred to Alice Springs (Lat. 24°S, Long. 134°E), from Embleton (1973) and McElhinny (1973). Arrows at top of (D) and (E) indicate position of the present shelf edge. The global sea level curve shown in (E) is from Vail et al. (1977) and the scale is relative only.

ticular the USSR (Hallam, 1977). Any differences between the Australian curve (Fig. 4D) and the global curve in the period 0 to 200 Ma ago point to independent vertical motion of Australia, and in the period 200-570 Ma ago to relative vertical motion between Australia on the one hand and North America and USSR on the other.

In the period 200 to 0 Ma ago, a gradual rise in the Australian curve is punctuated by a sharp peak at about 100 Ma ago, whereas the global curve rises gradually to a broad peak at 65 Ma ago and thereafter falls to a minimum at the present day. The critical differences are the sharp Australian peak at about 100 Ma ago versus the broad global peak at about 65 Ma ago and the gradually rising Australian curve from 65 Ma ago to the present versus the falling global curve. The mismatch by 35 Ma of the Cretaceous sea-level peaks is interpreted as reflecting a local event in Australia: the subsidence of the continental interior adjacent to an active convergent zone, manifested by contemporaneous magmatic activity along the eastern margin, including New Zealand (Veevers and Evans, 1973, 1975); and the gradual rise of the Australian curve is interpreted as the dual effect of the marginal subsidence that followed continental breakup by plate divergence on the western, southern, and eastern margins, and the subsidence of the continental interior (Arafura Sea and Gulf of Carpentaria) concomitant with plate convergence in New Guinea, manifested by contemporaneous magmatic activity. Equally noteworthy is the low sea level in Australia during the period 100 to 50 Ma ago when elsewhere sea level was high. This comparatively low stand of the sea in Australia corresponds with a period of reduced convergence, as manifested by an absence of plutonism, suggesting that whereas the plate interior may be depressed during convergence, it correspondingly rises during reduced convergence. This is what Johnson (1971, 1972) called the Haug effect: "times of orogeny are times of transgression of epicontinental seas on the continental interiors" and what Sloss and Speed (1974) called the submergent cratonic episode: "submergent cratons seem to coincide with times of active plate convergence expressed by obduction and subduction of the oceanic margins of cratons."

In the period 570 to 200 Ma ago, the global curve of Vail et al. (1977) becomes in effect a North American curve, and it rises from a minimum value of 570 Ma ago (beginning of the Cambrian) to peaks at 500 Ma ago (Early Ordovician), 400 Ma ago (Late Silurian-Early Devonian), and 335 Ma ago (Early Carboniferous), and falls off rapidly to minima at 325 Ma ago (early Carboniferous) and 265 Ma ago (Early Permian) on either side of a peak at 270 Ma ago (Early Permian). The Australian curve has a sharp peak at 540 Ma ago (early Middle Cambrian), a broad peak from 500 to 450 Ma ago (Early Ordovician), a low peak at 365 Ma ago (Late Devonian), a broad trough from 350 to 300 Ma ago (most of the Carboniferous), and then a slow rise to the Jurassic and later. Except in the period 300 to 200 Ma ago, over which they are grossly concordant, the curves are mismatched, most notably at 540, 430-400, and 350 to 330 Ma ago. The Australian highstand of sea level at 540 Ma ago is interpreted as a local marine transgression that followed the initiation by divergence of the Pacific and Tethyan margins, and the Australian lowstand of sea level at 350 to 330 Ma ago is possibly attributable to the broad uplift, by partial closure of a failed arm, of central Australia ("Alice Springs Orogeny") at this time.

A noteworthy correlation is found in Australia in the period 560 to 550 Ma ago between a rapid change from high to moderate paleolatitude and the wide-scale plate divergence, and both are explained as the result of a new regime of rapid seafloor spreading. The other correlation with paleolatitude is climatic: the occurrence of glaciation with high latitude in the Late Proterozoic and Late Carboniferous/Permian, and of aridity (as indicated by evaporites) with low paleolatitude in the Cambrian-Early Carboniferous and in the late Cenozoic.

Acknowledgments

This research was supported in part by the Australian Research Grants Committee.

References

Audley-Charles, M. G., J. R. Curray, and G. Evans, Location of major deltas, Geology, 5, 341-344, 1977.

Boeuf, M. G., and H. Doust,, Structure and development of the southern margin of Australia, Australian Petrol. Explor. Assoc. J., 16, 33-43, 1975.

Brown, C. M., P. E. Pieters, and G. P. Robinson, Stratigraphic and structural development of the Aure Trough and adjacent shelf and slope areas, Australian Petrol. Explor. Assoc. J., 15, 61-71, 1975.

Brown, D. A., K. S. W. Campbell, and K. A. W. Crook, The geological evolution of Australia and New Zealand, Pergamon, Oxford, 409 pp., 1968.

Carter, D. J., M. G. Audley-Charles, and A. J. Barber, Stratigraphic analysis of island arc-continental margin collision in eastern Indonesia, J. Geol. Soc. Lond., 132, 179-198, 1976.

Chamalaun, F. H., and A. E. Grady, The tectonic development of Timor: a new model and its implications for petroleum exploration, Australian Petrol. Explor. Assoc. J., 18, 102-108, 1978.

Cooper, J. A., Isotopic datings of the basement-cover boundaries within the Adelaide 'Geosyncline,' Interna. Aust. Geol. Conv. Abstr., Geol. Soc. Aust., 12, 1975.

Deighton, I., D. A. Falvey, and D. J. Taylor, Depositional environments and geotectonic framework: Southern Australian continental margin, Australian Petrol. Explor. Assoc. J., 76, 25-36, 1976.

Doutch, H. F., The Karumba Basin, northeastern Australia and southern New Guinea, BMR J. Aust. Geol. Geophys., 1, 131-140, 1976.

Embleton, B. J. J., The palaeolatitude of Australia through Phanerozoic time, J. Geol. Soc. Aust., 19, 475-482, 1973.

Exon, N. F., and J. B. Willcox, The geology and petroleum potential of the Exmouth Plateau area off Western Australia, Bull. Am. Assoc. Petrol. Geol. 62, 40-72, 1978.

Exon, N. F., and B. R. Senior, The Cretaceous of the Eromanga and Surat Basins, BMR J. Aust. Geol. Geophys., 1, 33-50, 1976.

Geological Society of Australia, Tectonic Map of Australia and New Guinea, 1:5,000,000, Sydney, 1971.

Hallam, A., Secular changes in marine inundation of USSR and North America through the Phanerozoic, Nature, 269, 769-772, 1977.

Harrington, H. J., K. L. Burns, and B. R. Thompson, Gambier-Beaconsfield and Gambier-Sorell fracture zones and the movement of plates in the Australia-New Zealand region, Nature, 245, 109-112, 1973.

Heckel, P. H., Recognition of ancient shallow marine environments, Soc. Econ. Palen. Miner., Spec. Publ. 16, 226-286, 1972.

Johnson, J. G., Timing and coordination of orogenic, epeirogenic, and eustatic effects, Geol. Soc. Am. Bull., 82, 3263-3298, 1971.

Johnson, J. G. Antler effect equals Haug effect, Geol. Soc. Am. Bull., 83, 2497-2498, 1972.

McElhinny, M. W., Palaeomagnetism and plate tectonics, Cambridge Univ. Press, 358 pp., 1973.

Packam, G. H., A speculative Phanerozoic history of the southwest Pacific, in P. J. Coleman, ed., The Western Pacific, Univ. W. Aust. Press, 369-388, 1973.

Page, R. W., Geochronology of igneous and metamorphic rocks in the New Guinea Highlands, Bur. Min. Resour. Aust. Bull., 162, 117 pp., 1976.

Powell, C. McA., and B. D. Johnson, Constraints on the Cenozoic position of Sundaland, Tectonophysics, 1979, in press.

Powell, C. McA., B. D. Johnson, and J. J. Veevers, A revised fit of East and West Gondwanaland, Tectonophysics, 1979, in press.

Quilty, P. G., Cenozoic sedimentation cycles in Western Australia, Geology, 5, 336-340, 1977.

Quilty, P. G., Mesozoic and Cenozoic history of Australia as it affects the Australia biota, in H. Cogger, ed., The arid zone in Australia, Aust. Museum Sesquicentenary, 1979, in press.

Scheibner, E., A plate tectonic model of the Palaeozoic tectonic history of New South Wales, J. Geol. Soc. Aust., 20, 405-426, 1973.

Sloss, L. L., and R. C. Speed, Relationships of cratonic and continental-margin tectonic episodes, Soc. Econ. Palen. Miner., Spec. Publ. 22, 98-119, 1974.

Sutherland, F. L., D. Stubbs, and D. C. Green, K-Ar ages of Cainozoic volcanic suites, Bowen-St. Lawrence hinterland, North Queensland, J. Geol. Soc. Aust., 24, 447-460, 1978.

Thompson, J. E., Continental drift and the geological history of Papua-New Guinea, Australian Petrol. Explor. Assoc. J., 12, 64-69, 1972.

Vail, P. R., R. M. Mitchum, and S. Thompson, Global cycles of relative changes of sea level, in Charles E. Payton, ed., Seismic stratigraphy - application to hydrocarbon exploration, Am. Assoc. Petrol. Geol. Mem. 26, 83-97, 1977.

Veevers, J. J. Early Phanerozoic events on and alongside the Australian-Antarctic platform, J. Geol. Soc. Aust., 23, 183-206, 1976.

Veevers, J. J., and D. Cotterill, Western margin of Australia: evolution of a rifted arch system, Geol. Soc. Am. Bull., 89, 337-355, 1978.

Veevers, J. J., and P. R. Evans, Sedimentary and magmatic events in Australia and the mechanism of world-wide Cretaceous transgressions, Nature Phys. Sci., 245, 33-36, 1973.

Veevers, J. J., and P. R. Evans, Late Palaeozoic and Mesozoic history of Australia, in K. S. W. Campbell, ed., Gondwana Geology, ANU Press, Canberra, 579-607, 1975.

Veevers, J. J., and M. W. McElhinny, The separation of Australia from other continents, Earth-Science Rev., 12, 139-159, 1976.

Veevers, J. J., and A. S. Rundle, Channel Country fluvial sands and associated facies of central-eastern Australia: modern analogues of Mesozoic desert sands of South America, Palaeogeography, Paleoclimatology, Palaeoecology, 26, 1-16, 1979.

Wellman, P., and I. McDougall, Potassium-argon ages on the Cainozoic volcanic rocks of New South Wales, J. Geol. Soc. Aust., 21, 247-272, 1974a.

Wellman, P., and I. McDougall, Cainozoic igneous activity in eastern Australia, Tectonophysics, 23, 49-65, 1974b.

GENESIS AND GEODYNAMIC EVOLUTION OF THE TAOUDENI CRATONIC BASIN
(UPPER PRECAMBRIAN AND PALEOZOIC), WESTERN AFRICA

G. Bronner, J. Roussel, R. Trompette

Laboratoire de Géologie dynamique, Université d'Aix-Marseille III,
Centre de St-Jérôme, 13397 Marseille Cedex 4, France et Laboratoire
associé au C.N.R.S. n° 132 "Etudes géologiques ouest-africaines"

N. Clauer

Centre de Sédimentologie et Géochimie de la surface (C.N.R.S.),
Institut de Géologie Université L. Pasteur, 1, rue Blessig,
67084 Strasbourg Cedex, France

Abstract. The Taoudeni basin is one of the
major structural units of the western African
craton. To the north and northwest, it is
bounded by the basement of the Reguibat shield.
The southern boundary is the Léo shield, the
Mauritanide fold belt occurs to the west, and
the so-called Pan-African fold belt is to the
east. During the period 1100-1000 to 300 Ma,
the geodynamic evolution of the Taoudeni basin
was controlled by two main factors:
1. An early regional Pan-African orogenic
   event circa 650 Ma - which gave elasticity
   again to the West African shield and craton
   which was stable since 1800 Ma. Before
   650 Ma, the subsidence rate was extremely
   low, 3-4 m/Ma, then after 650 Ma, it rose to
   15 m/Ma, a normal value for large cratonic
   basins.
2. The more local occurrence of heavy material,
   like ferruginous quartzite associated with
   aluminous gneisses, in the Lower Precambrian
   basement, which increased the density of the
   lower part of the crust and induced, during
   the 1100-1000 to 650 Ma period, important
   but local subsidence.

Introduction - Structural Framework

The Taoudeni basin is one of the major
structural units of the western African craton.
It covers about 2 million square km. Its
average altitude is +200 m above sea level and
the average depth of its basal surface is about
-1000 m (Fig. 1). Therefore, the sediments
have a thickness of 1000 to 1500 m and a volume
of about 2 to 3 million cubic km. It is
bounded by the Reguibat shield to the north and
northwest, by the Léo shield to the south, by
the Mauritanide fold belt to the west and by
the so-called Pan-African fold belt to the east.

Two major rock units occur in the basement
shields: a Lower Precambrian or Archean series
mainly in the western regions and a Middle
Precambrian series which outcrops over most of
the eastern zones.

The Pan-African fold belt (650 to 550 Ma)
includes the Pharuside fold belt and its extension into the Gourma region, as well as into
the Dahomeyide fold belt to the south. The
Mauritanide fold belt is of Hercynian age in
the north, but includes parts of the Caledonian
and Pan-African fold belts in the south. It is
divided into two branches in Gambia; the eastern one is probably connected with the Rokelide
Pan-African fold belt.

The Taoudeni Basin and its Basement

We summarize the geology of the basement
which may well influence the subsidence of the
Taoudeni basin and then we will describe
briefly the sedimentary cover.

1. The Basement of the Taoudeni Basin

The basement crops out to the northwest of
the basin in the Reguibat shield and to the
south in the Léo shield. Two large sequences
may be differentiated (Bessoles, 1977; Figs. 2
and 3):
A lower sequence of Lower Precambrian (or
Archaean) age, which was highly metamorphosed, migmatized and granitized about
2700 Ma ago. It also contains schists with
hyperaluminous and ferruginous horizons
(magnetite quartzites) that are widely
exposed in the western part of the Reguibat
shield. Granites, migmatites, anorthosites
and charnockites occur with these schists.
The charnockites are particularly well

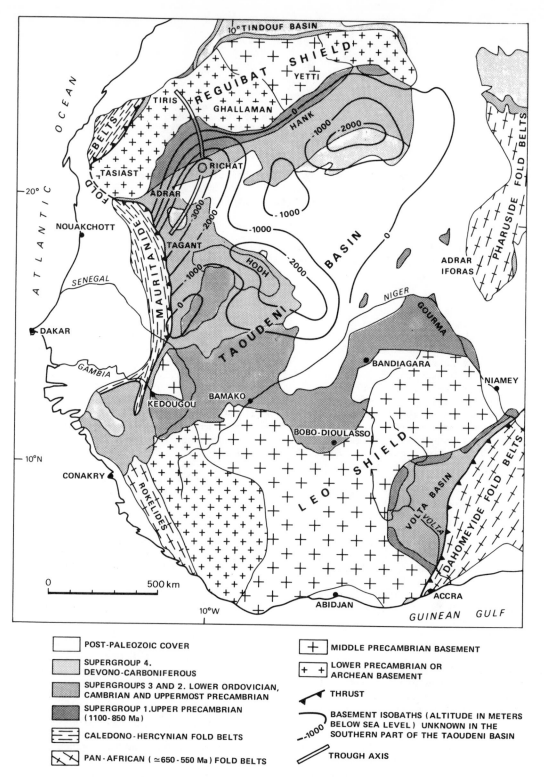

POST-PALEOZOIC COVER

SUPERGROUP 4.
DEVONO-CARBONIFEROUS

SUPERGROUPS 3 AND 2. LOWER ORDOVICIAN,
CAMBRIAN AND UPPERMOST PRECAMBRIAN

SUPERGROUP 1.UPPER PRECAMBRIAN
(1100-850 Ma)

CALEDONO-HERCYNIAN FOLD BELTS

PAN-AFRICAN (≈650-550 Ma) FOLD BELTS

MIDDLE PRECAMBRIAN BASEMENT

LOWER PRECAMBRIAN OR
ARCHEAN BASEMENT

THRUST

BASEMENT ISOBATHS (ALTITUDE IN METERS
BELOW SEA LEVEL) UNKNOWN IN THE
SOUTHERN PART OF THE TAOUDENI BASIN

TROUGH AXIS

Fig. 1. Schematic geologic map of Western Africa.

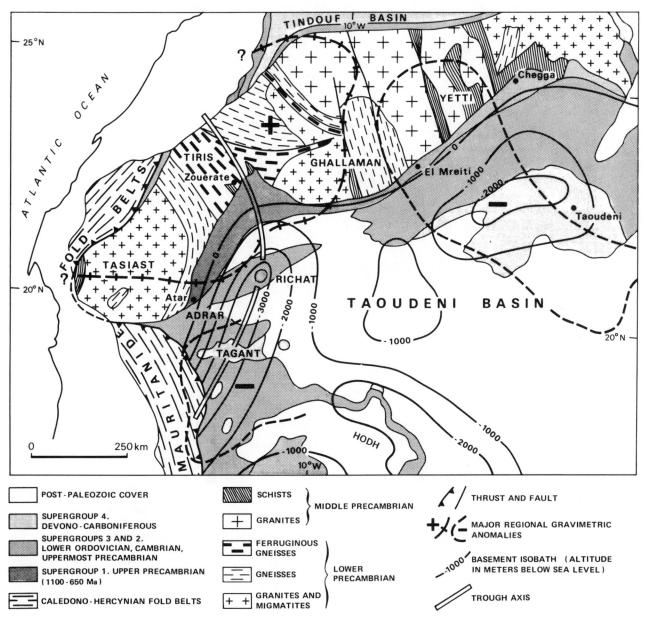

Fig. 2. Schematic geologic map of the northwestern border of
the Taoudeni basin and of its basement, the Reguibat shield.

developed in the eastern part of the Léo
shield, where the Lower Precambrian was
described by the name Liberian.

An upper sequence composed of Middle Precambrian, which was slightly metamorphosed
and granitized about 1800 Ma ago. These
granitizations are especially well represented in the Léo shield. The Precambrian
of the Léo shield was described under the
name Birrimian. It contains volcanic
spilitic-keratophyric rocks associated with
detrital sediments of a flyschoid facies.
The whole sequence is intruded by granitic

bodies which may be in a hypovolcanic
facies, especially in the Reguibat shield.

Let us now examine the distribution of these
two series 1100 to 1000 Ma ago, at the beginning of the sedimentation in the Taoudeni
basin. On both shields, the angle between the
actual erosion surface and the base of the
sedimentary formations is very low (1 to 2°).
This means that the present subcrop configuration of the two Precambrian series (Figs. 2
and 3) differs only slightly from what it was
1100 to 1000 Ma ago. Furthermore, on both
shields, the distribution of the two series is

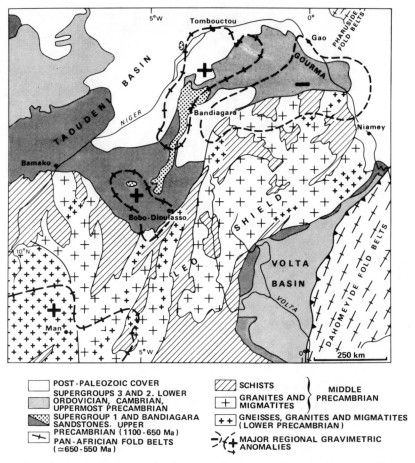

<!-- Legend -->
☐	POST-PALEOZOIC COVER
	SUPERGROUPS 3 AND 2. LOWER ORDOVICIAN, CAMBRIAN, UPPERMOST PRECAMBRIAN
	SUPERGROUP 1 AND BANDIAGARA SANDSTONES. UPPER PRECAMBRIAN (1100-650 Ma)
	PAN-AFRICAN FOLD BELTS (≃650-550 Ma)

SCHISTS	MIDDLE PRECAMBRIAN
GRANITES AND MIGMATITES	
GNEISSES, GRANITES AND MIGMATITES (LOWER PRECAMBRIAN)	
MAJOR REGIONAL GRAVIMETRIC ANOMALIES	

Fig. 3. Schematic geologic map of the southeastern part of the Taoudeni basin and of its basement, the Léo shield.

only slightly different in the central part when compared with the borders: both shields have a very old tectonic fabric which was completely stable since at least 1100 to 1000 Ma. Nevertheless, on the Reguibat shield, the upper sequence tends to be preferentially located at the border of the Taoudeni basin. This is the result of a slightly more active erosion in the axial zones, particularly at the northwestern extremity of this shield. All these observations support the ideas of Watson (1976) that the uplift and erosion of Lower and Middle Precambrian cratons occurred during a period immediately following the formation of the cratons.

The distribution of the two Precambrian series in the Reguibat shield (Bonner and Roussel, 1978) permits the identification of three zones:

A southwestern Lower Precambrian zone formed mainly by granites and migmatites in the southwest, gneisses that are very rich in magnetite quartzites in the Tiris region, and a mixture of gneisses and migmatites in the northeast. In this part of the shield,

evidence for a Middle Precambrian cover (schists on Fig. 2) is local and limited (west of the Tasiast region, southeast of the Tiris region and south of the Ghallaman region).

A central Middle Precambrian zone that is centered on the Yetti region and includes in essence granites with some preserved pieces of the cover, in the form of NNW-SSE oriented schists.

A northeastern zone with Lower Precambrian migmatites partially buried under a Middle Precambrian volcano-sedimentary cover.

Two interpretations can be offered to explain this distribution: variations in the sedimentation rate or in the postdepositional erosion. The presence of fragments of Middle Precambrian horizons in the southwestern zone suggests a more or less continuous Middle Precambrian cover. Its further erosion suggests a thinning of the crust which is also supported by gravimetry. The gravimetric map (Bouguer anomalies), of Mauritania (Rechenmann, 1971), indeed reflects the structure of the basement (Fig. 2). The positive regional anomaly, which appears in the southwestern Lower Precambrian part of the shield, can be explained by the presence of high density material at shallow depth. Such material formed at the bottom of the crust or the upper mantle, and its presence could explain why the crust is thinner (Louis, 1970). In contrast, the central and northeastern region of the shield is characterized by a negative regional anomaly which is roughly centered on the Middle Precambrian. This anomaly suggests a thickening of the crust caused by an especially thick Middle Precambrian cover. Such a very thick cover may be the result of active sedimentation that was possibly preserved in a graben structure. This regional anomaly shows, in detail, an alternation of parallel positive and negative local anomalies which may reflect horst and graben structures. The horsts would consist of Lower Precambrian material while the grabens contain Middle Precambrian rocks.

The distribution into two Precambrian zones is not quite so clear in the Léo shield (Fig. 3). Its western extremity includes the Lower Precambrian; the Middle Precambrian covers the central and eastern regions, but it often includes small fragments of Lower Precambrian migmatites. The gravimetric results of the Léo shield (Rechenmann, 1965) also reflect the contrast between the two sequences, but not as clearly as in the Reguibat shield. The Lower Precambrian basement is characterized, in the west, by a regional positive anomaly which could be related to a thinner crust. In the east of Man, this anomaly overlaps the Middle Precambrian. Therefore, the old Lower Precambrian basement probably extends to the east under a thin Middle Precambrian cover. Local positive anomalies, on the north of Man, for example, are probably related to the existence of high density Lower Precambrian material of granulitic facies (charnockites, magnetite quartzites, amphibolites, pyroxenites). On the east, the regional negative anomalies are diffuse and they are not shown on Figure 3. They extend roughly in a N 20° E direction and are closely related to the Middle Precambrian tectonics.

2. The Taoudeni Basin

We exclude from this study the description of the borders of the basin which were folded during orogenesis (Rokelide on the southwest, Pharuside and Gourma belt on the east during Pan-African time, and Mauritanide on the west during Caledono-Hercynian time). The basin is centered on the western African shield. Its sedimentary formations have ages ranging from the Upper Precambrian (ca. 1000 to 1100 Ma) to the Carboniferous. Stratigraphic sequences of these ages are largely exposed on the borders of the basin, but in the center they are mainly covered by a thin (100 m or less) Mesozoic and

Cenozoic veneer. The isobaths of the basement, which are only known in the north, the center, and the west of the basin, show a mean thickness of 1000 to 1500 m of sediment without any substantial deepening in the central part. Two large basins (Fig. 1), with about 2500 m of sediment, are formed in the Hodh region and around Taoudeni while a narrow asymmetric trough, oriented NNE-SSW, is clearly visible between the Richat and the Tagant regions. In the Adrar region this trough, which is the major element of the deep structure of the Taoudeni basin, could contain about 4000 m of sediments as estimated by aeromagnetic investigations; moreover, about 5000 to 6000 m of sediments were measured by field observations.

The Taoudeni basin has a simple structure with its youngest formations in the center. The sedimentation is typically regressive. The lithostratigraphic succession is divided into four discordant supergroups or sequences in Adrar of Mauritania (Trompette, 1973). Such a subdivision is valid for the whole basin.

(a) Supergroup 1 is the lower sequence which has an Upper Precambrian age that ranges from ca. 1000-1100 Ma to 650 Ma. In Adrar (Mauritania) and on the northwestern border of the basin, this supergroup contains an alternation of shales, siltstones, and often dolomitic carbonates with stromatolites. This sequence is sandwiched between sandy detrital beds (Fig. 4). The clastics came from NNE. Geochronological studies done in Adrar permitted an evaluation of the sedimentation rate, which is roughly equal to the rate of subsidence (Clauer, 1976). We will use the subsidence rate instead of the sedimentation rate because the thickness of the stratigraphic column only represents a minimal value after subtraction of the erosional effects. In Adrar (Mauritania) the subsidence rate (Fig. 5 and Table 1) is 4 m/Ma for the entire Supergroup 1 and 3 m/Ma for the intermediate carbonate section. It is therefore roughly independent of the facies.

Supergroup 1 maintains the same lithostratigraphy as in Adrar all the way to the northwestern border of the Taoudeni basin. Its thickness, however, varies greatly (Fig. 4): from 0 m in the south of Adrar to 3500 m in the Richat region with intermediate values at El Mreïti (200 m), at Chegga (800 m) and north of Atar (1450 m). As a consequence of this variation, the outcrop width of Supergroup 1 changes. The maximum width, which is located between the Tiris and the Richat regions, corresponds to the greatest thickness (3500 m). Such a thickening of Supergroup 1 suggests the prolongation of the Tagant-Richat trough, which north of the Richat region may curve towards the NNW.

The northerly continuation of the Tagant-Richat trough toward the Tiris region does not appear clearly on the schematic drawing of the

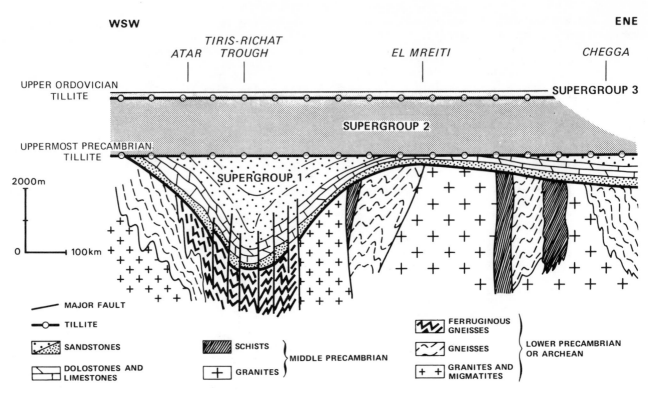

WSW   ENE

TIRIS-RICHAT
ATAR   TROUGH   EL MREITI   CHEGGA

UPPER ORDOVICIAN
TILLITE   SUPERGROUP 3

SUPERGROUP 2

UPPERMOST PRECAMBRIAN
TILLITE

2000m   SUPERGROUP 1

0   100km

—— MAJOR FAULT
—o— TILLITE

SANDSTONES   SCHISTS   FERRUGINOUS
                                              GNEISSES
                                         } MIDDLE PRECAMBRIAN   GNEISSES   LOWER PRECAMBRIAN
DOLOSTONES AND   OR ARCHEAN
LIMESTONES   + GRANITES   + + GRANITES AND
                                              + + MIGMATITES

Fig. 4. Schematic section of the northwestern border of the Taoudeni basin.

basement isobaths, but is in good agreement
with the curvature observed in the Lower Precambrian gneisses which crop out to the west of
Zouerate (Fig. 2). This is also supported by
the gravimetric anomalies (Crenn et al., 1962).
The bottom of Supergroup 1, which fills up the
trough, is cut by a succession of faults in the
Zouerate region. These faults were active
during sedimentation and indicate a succession
of horsts and grabens (Fig. 4) which are
located on the axis of the trough. These horst
and grabens appear to be the result of the
relative instability of the basement at the
beginning of the deposition of Supergroup 1.
It seems that the trough continues on the other
side of the Reguibat shield into the southeastern border of the Tindouf basin, where an
equivalent of the Taoudeni Supergroup 1
(Fig. 4) occurs.

The thickness variations observed in the
Supergroup 1 sequence reflect variations of the
subsidence rates (Fig. 6 and Table 1), which
are 4 m/Ma in the Adrar region, 10 m/1000 Ma in
the Richat region, 0.5 m/Ma at El Mreïti and
finally 2.5 m/Ma at Chegga.

We have less information in the southern
part of the basin. Nevertheless, near Bobo
Dioulasso, Supergroup 1 extends over the Léo
shield in a southerly direction. This onlap-
ping sequence with its principal axis located
to the southeast of the Niger river, could
relate to the thickening of the stratigraphic

succession which reaches 2000 m to the north of
Bobo Dioulasso and even more to the north of
Bandiagara. The Bandiagara continental sandstones (Fig. 3), deposited by streams flowing
from the SSE, become much thicker towards the
NNE (Bertrand-Sarfati and Moussine-Pouchkine,
1977) and may represent a channel system filled
up by Supergroup 1 deepening to the NNE.

The dominant gravimetric structure of this
zone is a double regional positive anomaly with
a NNE-SSW trend which is nearly superimposed on
the Bandiagara sandstone channels. This coincidence suggests a relation between the channel
system and the anomaly. Moreover, the remark-
able homogeneity of this anomaly with an
extremely small gravimetric continuity index
(Crenn et al., 1962), may indicate a sedimentary cover of at least 4 km to the north of
Bandiagara.

The connection of the Bandiagara channel
system with the Gourma subsidence zone, which
has been affected by the Pan-African orogeny,
is still poorly known. Finally, it is not
possible to estimate the subsidence rate of
Supergroup 1 on the southern border of the
Taoudeni basin because we lack precise thickness measurements and chronological data.

(b) Supergroup 2 rests unconformably on
Supergroup 1 or else it onlaps the basement.
Its age ranges between ca. 650 Ma and Lower
Ordovician. It starts with a triad over the
entire Taoudeni basin: tillite-baritic car-

TABLE 1. SUBSIDENCE RATES DURING THE UPPER PRECAMBRIAN AND THE PALEOZOIC IN THE TAOUDENI BASIN

	ADRAR OF MAURITANIA			OTHER AREAS
	Geochronological Data		Thickness (m) and Subsidence Rate (m/Ma)	Subsidence Rate (m/Ma)
Supergroups 1 + 2 + 3 + 4	Upper Devonian to Upper Precambrian (1100-1000 Ma)	650 Ma	3 350 5	
Supergroup 4	Upper Devonian to Upper Siegenien-Emsien	30 Ma	480 16	Northern Part of Taoudeni 11
Supergroup 3	Ludlow to Ashgill-Caradoc	40 Ma	180 4.5	
Supergroup 2	Lower Ordovician to Upper-most Precambrian (650 Ma)	100 Ma	1 250 12.5	Southern Part of Taoudeni 7-8
Supergroup 1	Upper Precambrian	350 Ma	1 400 4	see Fig. 6

bonates-siliceous rocks. This triad is followed by a green clayey formation which is sometimes replaced by greywackes. A red, mainly continental, clayey and sandy horizon, or a sandstone, forms the upper part of this sequence. It ends on the northwestern border, especially in Adrar (Mauritania), with scolithic sandstones containing a few faunas of inarticulated Brachiopods which are characteristic of the boundary between Cambrian and Ordovician. Supergroup 2 maintains a monotonous facies in the entire basin. In Adrar (Mauritania) and on the northwestern border of the basin, the subsidence rate is about 15 m/Ma (Fig. 5 and Table 1). But this rate decreases towards the southeast and on the southern border where it reaches only 7 to 8 m/Ma.

On the northwestern border, the sediments of the basal level of Supergroup 2, which are of glacial origin, were derived from the NNE while the intermediate and the lower parts contain material coming from SSE. It seems that these directions are consistent for the entire basin. The reversal of sources of clastic material is related to an early Pan-African orogenic event (ca. 650 Ma).

(c) Supergroup 3 rests with a disconformity on Supergroup 2 and contains two formations which have a very similar facies over the whole basin. The lowest horizons are of glacial origin, are of Late Ordovician age, and consist essentially of sandstones. In the Adrar region, the upper shales contain Silurian graptolites and have all the characteristics of a condensed sequence. The subsidence rate is 4 to 5 m/Ma in Adrar (Mauritania) (Fig. 5 and Table 1) and remains roughly consistent throughout the entire basin.

(d) Supergroup 4, of Devonian and Carboniferous age, rests also with an unconformity on Supergroup 3. It is entirely exposed in the Taoudeni region where it starts with continental sandstones of Siegenian and Emsian age. These sandstones are followed by carbonates and clayey sediments ranging in age from Devonian to Carboniferous. Finally, Upper Carboniferous sandstones and red clays with continental limestones represent the top of the sequence. The subsidence rate is ca. 16 m/Ma for the Devonian in Adrar (Fig. 5 and Table 1), whereas it is around 11 m/Ma for the Devonian and Carboniferous in the Taoudeni region (Villemur, 1967).

Interpretations and Reconstruction of the Geodynamic Evolution of the Taoudeni Basin

Two different sequences can be distinguished in the cover of the Taoudeni basin in terms of facies and subsidence rate:
A lower sequence containing Supergroup 1, which is characterized by a generally low but highly variable subsidence rate (from 0.5 to 10 m/Ma), by different facies and variable directions of sediment transport;
An upper sequence including Supergroups 2, 3, and 4, which is characterized by very homogeneous facies, relatively constant sub-

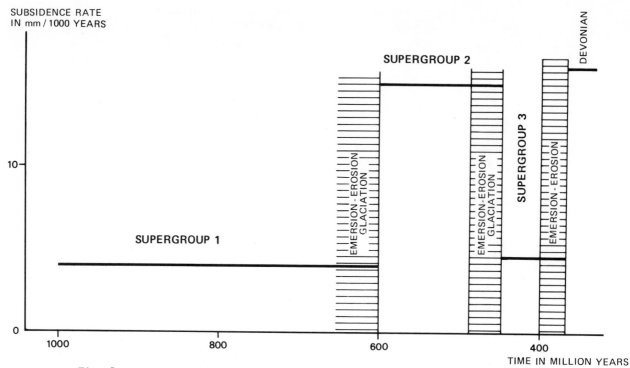

Fig. 5. Variations of the sibsidence rate during the Upper Precambrian and the Paleozoic in Adrar of Mauritania.

sidence rates (ranging between 4.5 and 16 m/Ma) and roughly identical sediment transport directions over the whole basin.

1. The Lower Sequence. The Influence of the Basement on Sedimentation

It seems that the distribution of Lower and Middle Precambrian formations within the basement have no influence on the sedimentation of Supergroup 1 (facies, thickness). However, the Tiris-Richat-Tagant trough, in the western part of the Reguibat shield, seems to be closely related, as its northern extremity, to the Lower Precambrian gneisses which are rich in interbedded iron quartzites (Figs. 2 and 4). We saw that the SSE-NNW orientation of this trough, in the northern part of the Richat region, coincides well with the structure of the Tiris gneisses. This coincidence permits

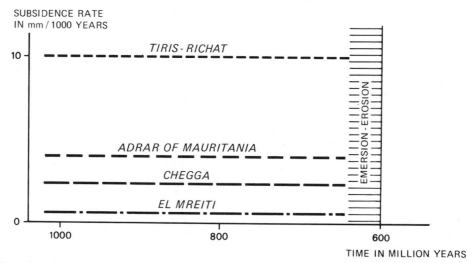

Fig. 6. Geographical variations of the subsidence rate of Supergroup 1 (Upper Precambrian) on the northwestern border of the Taoudeni basin.

the conclusion that the iron-rich gneisses are responsible for the subsidence of the Tiris-Richat-Tagant trough, an aulacogen-like structure. These iron-rich gneisses, which are exposed near Zouerate on the Reguibat shield, are associated with a strong local positive gravity anomaly. This anomaly has the characteristics of a 12 km thick heavy nucleus with a $0.1 \ g/cm^3$ mass excess with regard to the surrounding granites and migmatites. The extension of this dense mass under the cover of the Taoudeni basin causes the individualization of the Tiris-Richat-Tagant trough during deposition of Supergroup 1.

The subsidence appears to be especially high during deposition of the upper detrital part of Supergroup 1 (Fig. 4). The curvature of the trough, in the southern part of the Richat region towards SSW, agrees with what we know about the major structural directions in the basement. These directions are oriented NNE-SSW in the Kedougou window (Fig. 1).

The Bandiagara channel system could also be related to the presence of heavy masses in the Lower Precambrian basement, in the southern region of the basin. This hypothesis is based on two observations (Fig. 3):

the coincidence, discussed earlier, between the outcrops of the Bandiagara sandstones and the axis determined by the two positive anomalies;

the similar orientation of that axis with the anomalies and the major structural direction of the Léo shield.

The presence of Lower Precambrian material and the absence of any Middle Precambrian cover seem to be directly responsible for the positive anomaly and do not relate to sedimentation (Louis, 1970). However, the reason could be found elsewhere. Indeed, a series of positive anomalies extends the Bandiagara anomaly towards NNE. Near Adrar des Iforas, these join other positive anomalies which outline the eastern border of the craton. The positive gravimetric axis, which is over 1000 km long, could underline a major deep suture with, for example, intrusions of basic or ultrabasic material. These intrusions could have caused and controlled the subsidence in the Bobo-Dioulasso-Bandiagara region. However, such sutures have not been described in the Léo shield.

If one excepts the two Tiris-Richat-Tagant and Bandiagara troughs, Supergroup 1 looks like a thin veneer of sediments with a mean thickness of 1 to 1.5 km. This gives a subsidence rate of ca. 3 to 4 m/Ma lower than those generally found for cratonic areas, which range between 10 and 20 m/Ma (Perrodon, 1972; Ellenberger, 1976). Such a small rate of subsidence reflects the high stability of the western African craton during this period. This craton looks like a very flat platform where different

more or less independent basins appear. The Adrar sequence of Mauritania is often chosen as the type-section of Supergroup 1, because of its proximity to the depression, but it provides a poor picture of the average sedimentation of Supergroup 1 in the Taoudeni basin.

2. The Upper Sequence. The Structural Formation of the Western African Craton

The upper sequences (Supergroups 2, 3, and 4) are very homogeneous. The different formations, especially in Supergroups 2 and 3 which cover wide areas, can be found everywhere in the basin (Fig. 1). The subsidence rates of Supergroups 2 and 4 range from 7-8 m/Ma to 16 m/Ma and have thus the same values as those found in many large cratonic basins.

The very low subsidence rate of Supergroup 3 can be explained by the isostatic uplift of the basement. This uplift follows the Upper Ordovician glaciation and explains the thinness of the Silurian in the entire basin. The influence of the Upper Precambrian glaciation on the subsidence rate of Supergroup 2 cannot be estimated without more precise geochronological data.

A subsidence axis, oriented WSW-ENE to SW-NE and underlined by the basement isobaths as well as by the exposure of Supergroup 4, has formed in the Taoudeni basin from the beginning of the sedimentation of the upper sequence. This large but shallow depression seems asymmetrical with a thicker Supergroup 2 on the northwestern border where Supergroup 4 outcrops.

The Reguibat and Léo shields have roughly the same orientation as this depression. Their formation as shields probably occurred at the same period, that is during the interval between the deposition of Supergroups 1 and 2.

The influence of the Tiris-Richat-Tagant trough, which was active during the sedimentation of Supergroup 1, on the sedimentation of the upper sequence seems very weak. This is shown by the same relative thickness of Supergroup 2 on the entire northwestern border of the basin (Fig. 4).

Conclusion

During the period 1100-1000 to 300 Ma, the geodynamic evolution of the Taoudeni basin, and more especially the variations of the subsidence rate, was controlled by two main factors:

an early regional Pan-African orogenic event circa 650 Ma - which gave elasticity again to the West African shield and craton which was stable since 1800 Ma. Before 650 Ma, the subsidence rate was extremely low, 3-4 m/Ma, then after 650 Ma, it rose to 15 m/Ma, a normal value for large cratonic basins;

the more local occurrence of heavy material, like ferruginous quartzite associated with aluminous gneisses, in the Lower Precambrian basement - which increased the density of the lower part of the crust (Bott, 1976) and induced, during the 1100-1000 to 650 Ma period, important but local subsidence.

Acknowledgments

Our sincere thanks to F. C. Barbis of Ohio State University, Columbus, Ohio (U.S.A.) who kindly polished the English translation.

References

Bertrand-Sarfati, J., and A. Moussine-Pouchkine, Les grès de Bandiagara: Sédimentologie et paléocourants. Rapp., Centre Géol. Géophys., Montpellier, Fr., 67-70, 1977.

Bessoles, B., Géologie de l'Afrique. Le craton ouest-africain, Mém. B.R.G.M., Paris, 88, 402 p., 1977.

Blanchot, A., J. P. Dumas, and A. Papon, Carte géologique de la partie meridionale de l'Afrique de l'Ouest a 1/2000 000, Publ. inéd., B.R.G.M., Paris, 5 p., 1973.

Bott, M. H. P., Mechanisms of basin subsidence, an introductory review, Tectonophysics, 36, 1-4, 1976.

Bronner, G., and J. Roussel, Données gravimétriques et structure de la dorsale Reguibat (Mauritanie), 6e Réun. ann. Sc. Terre, Fr., Orsay, 72, 1978.

Choubert, G., and A. Faure-Muret, Carte tectonique de l'Afrique, Fr., Unesco, Paris, 1968.

Clauer, N., Géochimie isotopique du strontium des milieux sédimentaires, Application a la géochronologie de la couverture du craton ouest-africain, Mém. Sci. Géol., Strasbourg, Fr., 45, 256 p., 1976.

Crenn, Y., Cl. Blot, J. Metzger, and J. Rechenmann, Mesures gravimétriques et magnétiques en Afrique occidentale de 1956 à 1958, Cah. O.R.S.T.O.M., Sér. Géophys., Paris, 3, 45 p., 1962.

Crenn, Y., and J. Rechenmann, Mesures gravimétriques et magnétiques au Senegal et en Mauritanie occidentale, Cah. O.R.S.T.O.M., Sér. Géophys., Paris, 6, 59 p., 1965.

Deynoux, M., Essai de synthèse stratigraphique du bassin de Taoudeni (Précambrien supérieur et Paléozoïque d'Afrique occidentale), Trav. Lab. Sci. Terre, St-Jérôme, Marseille, Fr., B, 3, 71 p., 1971.

Ellenberger, F., Epirogènese et décratonisation, Bull. B.R.G.M., Paris, I, 4, 357-382, 1976.

Louis, P., Contribution géophysique a la connaissance géologique du bassin du Lac Tchad., Mém. O.R.S.T.O.M., Paris, 42, 311 p., 1970.

Marchand, J., J. Sougy, G. Rocci, J. P. Caron, M. Deschamps, B. Simon, M. Deynoux, C. Tempier, and R. Trompette, Etude photogéologique de la partie orientale de la dorsale Réguibat et de sa couverture sud (Mauritanie), Tome 1: synthèse géologique, Trav. Lab. Sci. Terre, St-Jérôme, Marseille, Fr., X, 11, 167 p., 1971.

Marchand, J., R. Trompette, and J. Sougy, Etude photogéologique de la région El Mreïti-Mejahouda-Agaraktem (Mauritanie), Trav. Sci. Terre, St-Jérôme, Marseille, Fr., X, 21, 28 p., 1972.

Perrodon, A., Esquisse d'une géologie dynamique des bassins sédimentaires, Sci. Terre, Nancy, Fr. 4, 301-328, 1969.

Perrodon, A., Réflexions sur la comparaison de quelques vitesses de phénomènes géologiques, C.R.. Som. Soc. géol. Fr., 2, 50-52, 1972.

Rechenmann, J., Mesures gravimétrique et magnétique en Côte d'Ivoire, Haute-Volta et Mali méridional, Cah. O.R.S.T.O.M., sér. Géophys., Paris, 5, 43 p., 1965.

Rechenmann, J., Cartes gravimétriques et magnétiques du Nord Mauritanie, 2 cartes (Centre Geophysique de M'Bour, 1961) et notice explicative n°46, O.R.S.T.O.M., Paris, 4 p., 1971.

Trompette, R., Le Précambrien superieur et le Paléozoïque inferieur de l'Adrar de Mauritanie (Bordure occidentale du bassin de Taoudeni, Afrique de l'Ouest). Un exemple de sédimentation du craton, Etude stratigraphique et sédimentologique, Trav. Lab. Sci. Terre, St-Jérôme, Marseille, Fr., B, 7, 702 p., 1973.

Villemur, J. R., Reconnaissance géologique et structurale du bassin de Taoudeni, Mém. B.R.G.M., Paris, 51, 172 p., 1967.

Watson, J. V., Vertical movements in Proterozoic structural provinces, Phil. Trans. R. Soc. London, A. 280, 629-640, 1976.

SUBSIDENCE OF THE PARIS BASIN FROM THE LIAS TO THE LATE CRETACEOUS

Cl. Mégnien

B. R. G. M., B.P. 6009, 45018 Orleans Cedex, France

Ch. Pomerol

Université P. et M. Curie, Laboratoire de Géologie des Bassins Sédimentaires
4 Place Jussieu, 75230 Paris Cedex 05, France

Abstract. During the Jurassic and Cretaceous, the pole of subsidence of the Paris Basin
migrated from the NNE to the SSW. It seems that
the subsidence is not a decreasing exponential
function of time but is more likely to be due in
part to the flexing of the lithosphere under
the weight of sediments.

Discussion

The Paris Basin is an intracratonic basin
600 km in diameter that rests on ancient mountain masses (Massif Armoricain, Massif Central,
Vosges, Massif Ardenno-Rhenan). To the northwest it opens widely toward the Belgian Basin,
the south end of the North Sea, and to the west
toward the English Channel into which its deep
structures extend.

Maximum subsidence is centered around the
region of Meaux (Brie meldoise) where the pre-
Permian basement is more than 3,000 m deep. The
ratio of depth to diameter is about 5/1,000
(Pomerol, 1977).

It is possible to suppose that subsidence of
the Paris Basin began during upper Triassic
Keuper time since, from the beginning of the
Trias up to the base of the Keuper, the area of
the basin was only a part of the German Basin
and thus remained widely open to the east;
moreover, during the Cenozoic, the Paris Basin
no longer acted like an independent marine basin
as the deposits are strongly influenced by local
conditions due to the particular tectonogenesis
of this epoch (Cavelier and Pomerol, 1978).

Therefore, this study is limited to the
Jurassic and Cretaceous periods; it is based on
the geologic interpretation of a large number of
deep boreholes and on isopach maps of the main
formations, achieved during an ongoing synthesis
of the Paris Basin.

Figure 1 represents the successive positions
occupied by the pole of maximum subsidence from
Lias to the Late Cretaceous. It is evident that

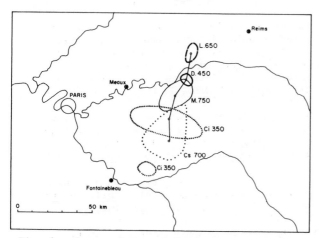

Fig. 1. Successive positions occupied by the
pole of maximum subsidence from the Lias to the
late Cretaceous in the Paris Basin. Numbers
indicate the maximum subsidence in meters.
L = Lias; D = Dogger, M = Malm; Ci = Lower
Cretaceous; Cs = Upper Cretaceous.

Table 1 MAXIMUM SUBSIDENCE OF THE PARIS BASIN
FROM THE LIAS TO THE LATE CRETACEOUS

Age from Phanerozoic Time Scale (1971). Other
dates proposed since 1971 do not alter the
graphs (Figs. 2 and 3) in a significant manner.

EPOCHS	T DURATION	E THICKNESS	E/T m/Ma
UPPER CRETACEOUS	35 Ma	700 m	20
LOWER CRETACEOUS	40 Ma	350 m	9
MALM	15 Ma	750 m	50
DOGGER	20 Ma	450 m	23
LIAS	25 Ma	650 m	26

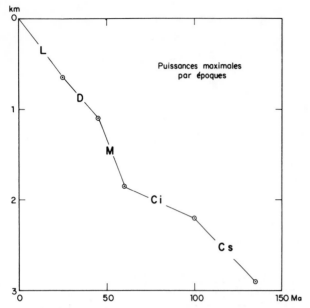

Fig. 2 Cumulated thicknesses of sediments from Lias to Upper Cretaceous in the Paris Basin (for the symbols, see Fig. 1).

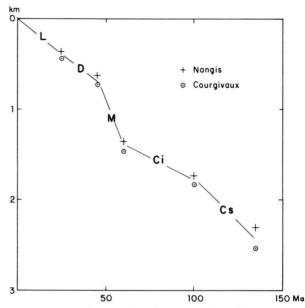

Fig. 3. Cumulated thicknesses of sediments from Lias to Upper Cretaceous in the boreholes at Nangis and Courgivaux (Brie, Paris Basin) (for the symbols, see Fig. 1).

this pole migrated about 60 km in 135 Ma from the NNE to the SSW.

In order to evaluate the maximum subsidence of the basin, it was considered that, for each epoch, subsidence is represented by the maximum thickness of the sediments (Table 1); these thicknesses were then added, starting from the base of the Lias (Fig. 2).

The graph of Figure 2 indicates a total maximum subsidence of 2,900 m in 135 Ma, that is, an average of 21.5 m per million years (21.5 m/Ma), which is the same order of magnitude as that for other intracratonic basins (Ellenberger, 1976; Pomerol, 1977). The average value applied to the Lias, Dogger, and the late Cretaceous; on the contrary, during deposition of the Malm, subsidence appears to have been much greater (50 m/Ma), whereas it is much less during the early Cretaceous (9 m/Ma). For the latter period it is, however, important to note that a continental episode of long duration affects the base of the formation (Wealdian, Barremian) and that the subsidence essentially occurred in the Albo-Aptian.

While the Tertiary is not considered in this analysis, it is also possible to observe a reduction of subsidence during this epoch, which is connected with continentalization of the basin.

One could object that these conclusions are valid only for the maximum thicknesses and not for the values measured in a single borehole. The graphs of Figure 3, established respectively for the boreholes at Nangis and Courgivaux, show the same tendencies.

We may conclude that the center of the Paris Basin during the Jurassic and Cretaceous does not, in any case, subside as a decreasing

exponential function of time, like those which characterize the basins of the continental margins (Foucher and Le Pichon, 1972).

In these conditions, subsidence would be principally due to the flexing of the lithosphere under the weight of sediments, accompanied by a slow migration of the pole of subsidence (Walcott, 1972) rather than a thermal contraction of the lithosphere (Poulet, 1977).

References

Cavelier, C., and Ch. Pomerol, Chronologie et interprétation des évènements tectoniques cénozoïques dans le Bassin de Paris: Bull. Soc. Géol. Fr., 1978, in press.

Ellenberger, F., Epirogenese et décratonisation: Bull. B.R.G.M., 2, Sec. 1, 357-382, 1976.

Foucher, J. P., and X. Le Pichon, Comments on thermal effects of the formation of continental margins by continental breakup, by N. Sleep, Geophys. J. Astron. Soc. 29, 43-46, 1972.

Poulet, M., Mécanismes de formation des basins sédimentaires de marge stable, Bull. Centre Rech. Explor. Prod., Elf Aquitaine, 1, 131-145, 1977.

Pomerol, Ch., Dynamique comparée de trois bassins epicontinentaux, Mer du Nord, Manche et Bassin de Paris, Bull. Centre Rech. Explor. Prod., Elf Aquitaine, 1, 233-256, 1977.

Walcott, R. I., Gravity, flexure and the growth of sedimentary basins at a continental edge, Geol. Soc. America Bull., 83, 1845-1848, 1972.

THE MICHIGAN BASIN

Norman H. Sleep

Department of Geophysics, Stanford University, Stanford, California 94305

L. L. Sloss

Department of Geological Sciences, Northwestern University
Evanston, Illinois 60201

Abstract.   The Michigan basin has been sampled extensively by boreholes. Phanerozoic
units thicken toward the center of the basin in
sub-circular isopach patterns. General concordance of isopach and facies patterns with
deeper-water facies occupying the basin interior
indicate subsidence of the basin concomitant
with deposition. Three regional unconformities,
corresponding to the boundaries of craton-wide
stratigraphic sequences (Sauk, Tippecanoe,
Kaskaskia) interrupt the Paleozoic stratigraphy
of the basin and represent basin-wide cessations
of subsidence followed by renewed downwarp.

The cause of basin subsidence is not yet well
understood; surficial loading is inadequate to
explain the magnitude of subsidence and at least
two heating events (Cambrian and pre-Middle
Ordovician) would be required if thermal contraction of the lithosphere is invoked. Deep
drilling has found no direct evidence of a post-
Cambrian thermal event.

Petroleum exploration and studies of borehole
samples will probably continue at a steady pace
in the Michigan basin. Integration of resulting
new data with published stratigraphic and sedi-
mentologic information will be a prerequisite to
the presentation of "hard" data in a form usable
by geophysicists.

Geological Setting

The basement complex in the Michigan basin
has been sampled by various drill holes and is
well exposed to the north and northwest of the
basin. A series of northwest-striking gravity
and magnetic anomalies defines the gross structure underlying the basin (Hinze et al., 1975).
A rift system related to the Keweenawan rocks in
the Upper Peninsula of Michigan and to the
Midcontinent gravity high is believed to be the
source of these anomalies (Chase and Gilmer,
1973; Hinze et al., 1975).

Phanerozoic sedimentary units in the Michigan
basin generally thicken toward the center of the
basin (Figs. 1, 2, and 3). This thickening,
general concordance of facies and isopachs and
the presence of deeper water facies near the
center of the basin indicate that subsidence of
the basin relative to its surroundings was concomitant with deposition. Middle Ordovician
through Early Carboniferous strata generally
exhibit this thickness and facies pattern while
Late Carboniferous detrital sediments exist but
are poorly represented. Jurassic redbeds occur
near the center of the basin but are unexposed
and nearly unstudied. Upper Cambrian and Lower
Ordovician units show modest thickening into the
basin and their facies exhibit some reflection
of basin geometry, but they are in part continuous with similar age units to the south and
west (Catacosinos, 1973). Some facies contacts
in this age interval cross cut the basin geometry along roughly east-west lines.

Recent Progress in Geology

A deep hole drilled near the center of the
Michigan basin penetrated 1600 m into Keweenawan
age redbeds and associated metamorphosed basic
igneous rocks (Sleep and Sloss, 1978). Paleomagnetic studies indicate that the redbeds are
Keweenawan in age (ca. 1000 Ma) and that an
igneous rock at 5224 m depth was metamorphosed
between about 500-800 Ma ago (Van der Voo and
Watts, 1978). Borehole gravity, in situ
stress, and heat flow were measured in the hole.
Strontium isotopes, clay minerals, and sedimentary and igneous petrology of the returned
samples were studied. No evidence was obtained
of any post-Early Ordovician thermal events in
the Late Precambrian samples.

Understanding of the sedimentology of the
Phanerozoic sediments is necessary to distinguish between periods of rapid deposition which
fill starved basins from periods of rapid subsidence. Much controversy in this regard exists

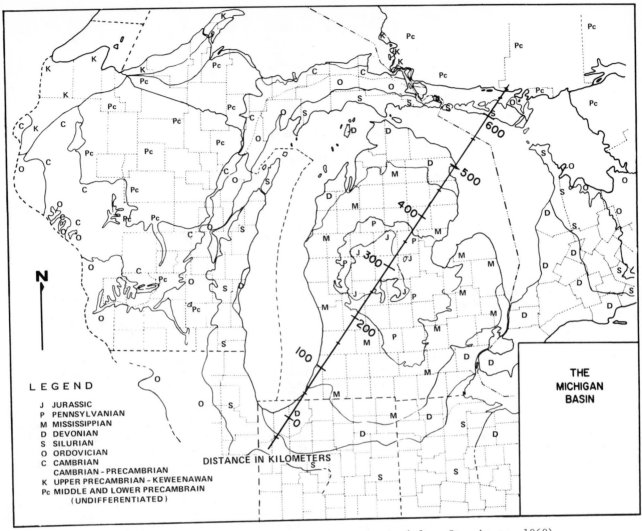

Fig. 1. Preglacial subcrop map of the Michigan basin (after Stonehouse, 1969).
The path of the cross-section in Figure 2 is roughly perpendicular to Keweenawan(?)
trends in the basement (by permission of the Michigan Basin Geological Survey).

with respect to Silurian salt deposition in the
basin. Michigan workers (Huh et al., 1977) have
generally contended that the Michigan basin was
starved during the deposition of "Niagaran"
reefs around the periphery of the basin and that
the salt deposits formed rapidly in Late Silurian time. Indiana workers (e.g., Droste and
Shaver, 1977) have noted that at least some of
the "Niagaran" reefs are Late Silurian and
propose that salt deposition, although episodic,
was concurrent with reef deposition.

Some idea of the water depths present in the
basin can be obtained from Silurian reefs which
have been extensively explored for oil in
Michigan and for quarry stone in Indiana and
Illinois. The Thornton reef near Chicago,
Illinois, on the flank of the basin shows about
100 m of relief (Pray, 1976). Pinnacle reefs in

the northern Lower Peninsula of Michigan are as
high as 180 m (Huh et al., 1977) although it is
possible that the parts of reefs may have grown
after some salt was deposited at their bases
(Mesolella et al., 1974).

Geodynamic Significance

The subsidence and tectonics of the Michigan
basin cannot be simply related to the kinematics
of plate tectonics. Significant dynamic questions which are not yet adequately answered
include:

1. The cause(s) of the subsidence of the
basin.
2. The relationship of continent- and
possibly world-wide unconformities to the

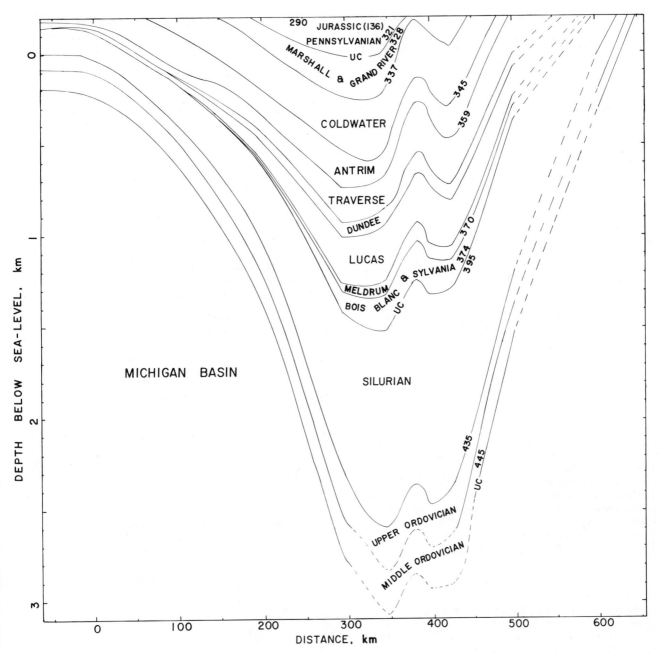

Fig. 2. Post-Sauk cross-section of the Michigan basin (after Sleep and Snell, 1976). Major unconformities (Sloss, 1963) are marked with U̲C̲ and the ages above and below given. Absolute ages are given where possible for formation boundaries. The anticline at 380 km is due to Late Paleozoic tectonism. Some minor structures were omitted.

mechanism of basin subsidence and to eustatic changes in sea level.

3. The relationship of basement structure to basin subsidence including controls of basin location, and the amount of differential subsidence which is taken up by basement faults.

4. The relationship of basin-subsidence history and large-scale tilting of continents, such as is indicated by subcrop patterns formed during major continent-wide erosional episodes.

The latter two subjects have not been investigated in the Michigan basin to any extent.

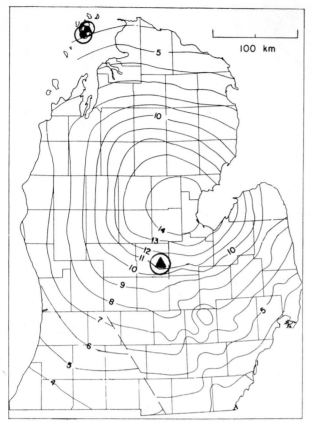

Fig. 3. Structural contours in kilofeet on the
Precambrian (modified after Hinze and Merritt,
1969). Note the gradual increase in depth to
basement toward the center of the basin. The
deep borehole discussed by Sleep and Sloss (1978)
is indicated by a triangle. Two boreholes on
Beaver Island which penetrated Precambrian are
indicated by circles (by permission of the
Michigan Basin Geological Society).

Recent Progress in Geodynamics

Because of its well-documented geology,
simple geometry, and quiescent tectonic history,
the Michigan basin has been selected for several
recent studies of basin subsidence. Sleep
(1971), Sleep and Snell (1976), and Haxby et al.
(1976) present variants of an hypothesis that
thermal contraction of the lithosphere to its
original state after a heating event causes
basin subsidence. Sleep and Snell (1976) and
Haxby et al. (1976) explicitly include the
mechanical properties of the lithosphere in
their calculations. Sleep and Snell (1976)
consider the lithosphere to be a visco-elastic
plate about 100 km thick and attribute later
erosion along the flanks of the basin to rebound
of peripheral areas of the basin which were
dragged down during early stages of basin subsidence. Haxby et al. (1976) model the litho-

sphere as a thinner (30 km thick) elastic slab
(the rest of the lithosphere is considered to
have little long-term strength) and find a
reasonable fit to basin geometry.

The major difficulties with the thermal contraction hypothesis in the Michigan basin are
that there is no evidence of an Early Ordovician
heating event, as would be required for the
basin to subside after that time, and that at
least two heating events are required to explain
the observed subsidence (Fig. 4, Sleep and
Sloss, 1978). The major arguments for the
thermal contraction hypothesis are that it
produces subsidence elsewhere on the earth (at
mid-ocean ridges and probably at Atlantic continental margins), and perhaps most compelling-
ly, that the hypothesis is well enough understood to be quantitatively modeled.

Loading of the lithosphere must occur prior
to subsidence according to the thermal contraction hypothesis and to the hypothesis by
McGinnis (1970) that attributes subsidence to
loading by dense intrusions in the lithosphere.
Sleep and Snell (1976) showed that loading of
the Michigan basin prior to post-Early Ordovician subsidence would not have been caused by
subaerial erosion during pre-subsidence uplift,
because no significant erosional degradation is
observed. Haxby et al. (1976) speculate that a
large intrusion of basin rocks crystallized
beneath the basin and then transformed to dense
eclogite. Direct or seismic information relevant to the cause of the loading is not available.

A small positive 1° x 1° free air anomaly, 50
to 100 mm/s^2 is centered on the Michigan basin
(Walcott, 1970). Walcott (1970), Sleep and

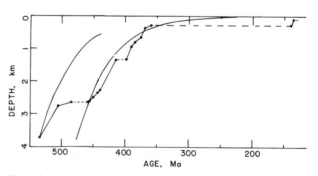

Fig. 4. The depths of strata in the deep
Michigan borehole plotted as a function of
absolute age. Time intervals missing due to
non-deposition or to erosion are indicated by
dashed lines. Theoretical 50 Ma exponentials,
expected if subsidence is due to thermal contraction of the lithosphere, are fit to the
present surface and to the bases of the Cambrian
and Middle Ordovician sections (solid curves).
One thermal event followed by cooling cannot
explain the observed subsidence (after Sleep
and Sloss, 1978).

Snell (1976) and Haxby et al. (1976) all attribute this anomaly to load-caused regional subsidence rather than to subsidence at the basin's center. Sleep and Snell (1976) compute a maximum viscosity of a 100 km thick lithosphere of 4×10^{25} poise compatible with the load inferred from this anomaly and note that glacial erosion in the Great Lakes basins and deposition on the Lower Michigan peninsula could partly cause the anomaly.

Haxby et al. (1976) use the same gravity anomaly to constrain the depth to the loading element which caused basin subsidence. If the basin subsided without being driven by a load, a large negative free air anomaly would result from the contrast in density between the sediments of the basin, 2.61 Mg/m^3 (Hinze et al., 1978), and the mantle which was displaced downward to make room for them. The load-causing subsidence would produce an equal and opposite positive anomaly if it were relatively shallow - that is, centered above about 50 km. The positive gravity anomaly created by a load in the asthenosphere would be so broad that a net negative anomaly would occur at the center of the basin.

Sequences and Unconformities in the Michigan Basin

Cratonic platform strata are divided into natural sequences by craton-wide unconformities (Sloss, 1963; Vail et al., 1977); such unconformities represent major deposititional hiatuses that decrease in time value from the margins to the centers of sedimentary basins, indicating that the degree of supra-baselevel uplift and the duration of subaerial exposure of basin interiors is less than that experienced by peripheral areas. Sloss and Speed (1974) recognize three tectonic modes of cratons; intra-sequence unconformities manifest one mode characterized by marine regression and a marked reduction in the relative vertical movements of basins and arches. This mode, termed emergent, appears to coincide with episodes of reduced continent-margin orogenesis, presumably reflecting low rates of sea-floor spreading and of plate convergence. Mid-Ordovician through Silurian and mid-Devonian through Early Carboniferous evolution of the Michigan basin is representative of submergent episodes during which sedimentary basins subside, largely by flexure, and there is wide marine transgression of continental interiors while plate convergence and consumption at continental margins is accelerated. The submergent mode in Michigan, as elsewhere, was replaced in Late Carboniferous time by higher-frequency vertical movements accompanied by high-angle faulting typical of oscillatory modes. Subsequent episodes of emergence have left a poor and deeply eroded remnant of Upper Carboniferous strata in the Michigan basin but faulting, erosion, and accompanying salt solution during this time profoundly affected basinal structure. A following submergent episode (mid-Jurassic and Cretaceous) may have imposed significant further load on the basin interior but erosion has left only minor Jurassic remnants. Interestingly, Pleistocene glacial sediments reach their greatest thickness ($\pm$ 200 m) in the interior of the basin, suggesting renewed response to glacial loading.

Continued Study

Because of the highly fragmented nature of published information on the Michigan basin, more compilation is needed to obtain a self-consistent set of isopachous and structural contour maps on the basin. The lithology and probably depth of deposition of the sediments are necessary to separate the load due to "driving forces" (Watts and Ryan, 1976) from the loads due to sediment thicknesses. Better time stratigraphic control would also help in comparing observed subsidence to that predicted from thermal contraction theory.

It may be expected that petroleum exploration and studies of borehole samples will continue at a steady pace in the near future. As in other areas published sedimentological and stratigraphic studies will continue to concern themselves with narrow stratigraphic intervals and often in limited geographic areas. Continued compilation will be necessary to present this information in forms usable by geophysicists.

References

Catacosinos, P. A., Cambrian lithostratigraphy of Michigan basin, Am. Assoc. Petrol. Geol. Bull., 32, 1417-1558, 1973.

Chase, C. G., and T. H. Gilmer, Precambrian plate tectonics: the mid-continent gravity high, Earth Planet. Sci. Letters, 21, 70-78, 1973.

Droste, J. B., and R. H. Shaver, Synchronization of depositon: Silurian reef-bearing rocks on Wabash Platform with cyclic evaporites of Mighigan basin, in J. H. Fisher, ed., Reefs and Evaporites - Concepts and Depositional Models, Am. Assoc. Petrol. Geol. Studies in Geology no. 5, 93-109, 1977.

Haxby, W. F., D. L. Turcotte, and J. M. Bird, Thermal and mechanical evolution of the Michigan basin, Tectonophysics, 36, 57-75, 1976.

Hinze, W. J., and D. W. Merritt, Basement rocks of the Michigan basin, in H. B. Stonehouse, ed., Studies of the Precambrian of the Michigan Basin, 1969 Ann. Field Excursion, Michigan Basin Geol. Soc., 28-59, 1969.

Hinze, W. J., R. L. Kellogg, and N. W. O'Hara, Geophysical studies of basement geology of southern peninsula of Michigan, Am. Assoc. Petrol. Geol. Bull., 59, 1562-1584, 1975.

Hinze, W. J., J. W. Bradley, and A. R. Brown, Gravimeter survey in the Michigan basin deep borehole, J. Geophys. Res., 83(B12), 5864-5868, 1978.

Huh, J. M., L. I. Briggs, and D. Gill, Depositional environments of pinnacle reefs, Niagara and Salina groups, northern shelf, Michigan basin, in J. H. Fisher, ed., Reefs and Evaporites - Concepts and Depositional Models, Am. Assoc. Petrol. Geol. Studies in Geology no. 5, 1-21, 1977.

McGinnis, L. D., Tectonics and the gravity field in the continental interior, J. Geophys. Res., 75(2), 317-331, 1970.

Mesolella, K. J., J. D. Robinson, L. M. McCormick, and A. R. Ormiston, Cyclic deposition of Silurian carbonates and evaporites in the Michigan basin, Am. Assoc. Petrol. Geol. Bull., 58, 34-62, 1974.

Pray, L. C., The Thornton reef (Silurian), northeastern Illinois, 1976 Revisitation, in Geol. Soc. Am. Guidebook for Field Trip, North Central Section, May 2, 1976.

Sleep, N. H., Thermal effects of the formation of Atlantic continental margins by continental break-up, Geophys. J. R. astr. Soc., 24, 325-350, 1971.

Sleep, N. H., Platform subsidence mechanisms and "eustatic" sea-level changes, Tectonophysics, 36, 45-56, 1976.

Sleep, N. H., and L. L. Sloss, A deep borehole in the Michigan basin, J. Geophys. Res., 83(B12), 5815-5831, 1978.

Sleep, N. H., and N. S. Snell, Thermal contraction and flexure of mid-continent and Atlantic marginal basins, Geophys. J. R. astr. Soc., 45, 125-154, 1976.

Sloss, L. L., Sequences in the cratonic interior of North America, Geol. Soc. Am. Bull., 74, 93-114, 1963.

Sloss, L. L., Synchrony of Phanerozoic sedimentary-tectonic events of the North American craton and the Russian Platform, Internat. Geol. Cong., 24th, Montreal, sec. 6, 24-32, 1972.

Sloss, L. L., and R. C. Speed, Relationships of cratonic and continental margin tectonic episodes, in W. R. Dickinson, ed., Tectonics and Sedimentation, SEPM Spec. Pub. 22, 98-119, 1974.

Stonehouse, H. B., ed., Map inside front cover, Studies of the Precambrian of the Michigan Basin, 1969 Ann. Field Excursion, Michigan Basin Geol. Soc., 1969.

Vail, P. R., Jr., R. M. Mitchum, Jr., R. G. Todd, J. M. Widmier, S. Thompson III, J. B. Sangree, J. N. Bubb, and W. G. Hatleid, Seismic stratigraphy and global changes of sea level, in Charles E. Payton, ed., Seismic Stratigraphy - Applications to Hydrocarbon Exploration, Am. Assoc. Petrol. Geol. Mem. 26, 49-212, 1977.

Van der Voo, R., and D. R. Watts, Paleomagnetic results from igneous and sedimentary rocks from the Michigan basin borehole, J. Geophys. Res., 83(B12), 5844-5848, 1978.

Walcott, R. I., Isostatic response to loading of the crust in Canada, Can. J. Earth Sci., 7, 716-727, 1970.

Watts, A. B., and W. B. F. Ryan, Flexure of the lithosphere and continental margin basins, Tectonophysics, 36, 25-44, 1976.

MODE AND MECHANISMS OF PLATEAU UPLIFTS

T. R. McGetchin[1], K. C. Burke[2], G. A. Thompson[3], and R. A. Young[4]

[1]Lunar and Planetary Institute, Houston, Texas 77058 (deceased)
[2]Department of Geological Sciences, State University of New York at Albany, Albany, New York 12222
[3]Department of Geophysics, Stanford University, Stanford, California 94305
[4]Department of Geological Sciences, State University of New York at Geneseo, Geneseo, New York 14454

Introduction

One of the keys to understanding the dynamics of plate interiors is the solution to the puzzle of plateau uplifts, especially those isolated from active mountain belts. These areas are of particular interest because they apparently involve deeply rooted processes and are independent of active subduction. Some of them are capped by young alkaline volcanics, while some are free of volcanics. Putorana in Siberia, the Adirondacks and Black Hills of North America and the Serra do Mar in Brazil are typical cases -- irregular areas 100-200 km in diameter, 1 km above their surroundings. Volcanic-capped uplifts are common in Africa (Ahaggar, Tibesti, Jos Plateau, Ngaoundere, and the Cameroon Zone), as are examples of the volcanic-free uplifts (Fouta Djallon, Angola, Adamawa, and high Veldt). The Colorado Plateau may be similar to volcanic-capped African uplifts.

This paper is based in large part on discussions which occurred as part of a conference on Plateau Uplift: Mode and Mechanism, which was held August 14-16, 1978, in Flagstaff, Arizona, was cosponsored by Working Group 7 of the Inter-Union Commission on Geodynamics and the Lunar and Planetary Institute, and was hosted by the U. S. Geological Survey. The 16 papers constituting the proceedings of this Conference appeared together in Tectonophysics in 1979.

Uplifts of the World

Since the term "uplift" implies upward structural displacement, there is commonly ambiguity and argument regarding what constitutes an uplift, in contrast to an area which is topographically high for a variety of other reasons. This is particularly true in areas where geological data are sparse. Following Holmes [1964], plateaus are defined as broad uplands of considerable elevation. Terrestrial plateaus occur both on the continents and on the ocean floor. Some are associated with plate margins, both convergent and divergent, and others are not, as Table 1 shows.

Broad areas of sea floor stand significantly above the mean depth; these are of considerable interest because the sea floor, in some sense at least, is simpler than the continents. Crough [1978] has developed a theory of ocean rises illustrated by the Hawaiian Swell. The lithosphere thins abruptly as it moves over hot spots and then thickens by slow cooling as it moves away, controlled by the same processes as the lithosphere moving away from a spreading ridge [Sclater and Francheteau, 1970; Sclater et al., 1971]. The rapid heating of the lithosphere, however, cannot be accomplished by conduction alone and may be caused by intrusion of dikes into the lithosphere. Plateaus of the sea floor may owe their elevation to any of several causes: some are volcanic piles (Manihiki and Ontong Java plateaus in the Pacific), some are believed to be continental fragments (Seychelles), and others, thermal perturbations (expanded mantle) [Jordan, 1975]. Problems occur when these features encounter active subduction zones. The floor of the Caribbean is believed to be occupied by such a buoyant residuum of ocean floor plateau [Burke, 1978].

Continental uplifts (Table 2), because of intense erosion rates, are by definition young, generally less than about 40 Ma old. (Hence the oldest plateaus on earth are to be found on the sea floor -- at substantial distances from the active spreading centers.) Continental plateaus are commonly, but not exclusively, associated with plate margins. An example of a plateau at divergent plate boundaries is in Ethiopia; those on convergent plate boundaries are in Tibet, Iran, and the Shillong and Altiplano plateaus. The latter are associated with active convergence.

An interesting and important insight on the question of the relative uplifts of large continental areas is provided by comparing data from the stratigraphic record on the percentage of continental areas flooded as a function of time

TABLE 1. Examples of the Types of Tectonic Settings of Plateaus of the World

Plate Margin		
Convergent	Divergent	Intraplate
Continental Plateaus		
Tibet	Ethiopia	Colorado Plateau
Iran		
Shillong		
Altiplano		
Oceanic Plateaus		
Caribbean	Iceland	Hawaii
	Galápagos	

and by use of hypsometric curves to infer the corresponding uplift. Use of the method [Bond, 1978] suggests extensive post-Miocene uplift of Africa, somewhat older (pre-Eocene) uplift of western North America, extensive uplift of Australia in Cretaceous time which had subsided substantially by Eocene time, and post-Cretaceous uplift of central Europe. The timing of such intraplate plateau uplifts in relation to other regional and global tectonics events is important. Burke [1978] has pointed out that the great African plateaus appear to begin in the early Tertiary, at a time when sea floor data elsewhere suggest that the African plate had come to rest with respect to the underlying convection pattern.

Continental uplifts also exist, however, which are not associated with currently active plate margins, namely, those lying within continents and remote from plate boundaries. The high plateau of central Mexico is certainly related to the rest of the American Cordillera; Urrutia-Fucugauchi [1978] suggests that the Mexican volcanic belt is the result of the subduction of the Cocos plate at the mid-America trench in middle to late Tertiary time and that subduction of the East Pacific Rise may occur in the near future. In this sense the Mexican highlands may be an important clue to the mid-Tertiary (pre-30 Ma) history of the western United States [Atwater, 1970; Lipman et al., 1972; Christiansen and Lipman, 1972; Coney, 1972].

The Rhenish shield, in western Germany and Luxemburg, is a relatively small (200 X 100 km), but interesting area because it has undergone some 150 m of uplift during the Quaternary. It is bounded on all sides by seismically active rift valleys and by young volcanism (Eifel area). An ambitious multidisciplinary investigation is underway; the current working model [Illies et al., 1979] suggests that the Rhenish shield is a fault-bounded structural block produced by northward displacement of the Alps during the late Tertiary and that the uplift and volcanism are due to shear heating resulting from slip at the base of the lithosphere. A comparison of the Rhenish and Fennoscandian shields [Theilen and

Meissner, 1979] shows that the Moho and asthenosphere are considerably deeper under Fennoscandia; it is suggested that Fennoscandia's uplift is due to its close association with the Atlantic, possibly involving mantle creep.

The Indian subcontinent [Kailasam, 1978] is flanked on the north by the Tibetan plateau, the largest and highest in the world -- this is the convergent plate boundary between the continental plates of India and Eurasia; the Shan plateau to the east is smaller, but also has experienced uplift over an extensive area. Four smaller, but substantial areas within peninsular India have undergone epeirogenic uplift during the Cenozoic -- the Deccan and Karnataka in south India and Chota Nagpur and Shillong plateaus in eastern and northeastern India. All are being investigated for both scientific and economic reasons; geodesy, seismicity, gravity, and volcanic geology suggest that these plateaus are active. A regional negative gravity anomaly is believed to indicate an unusually hot upper mantle and deeply rooted causes for the uplifts, such as plumes, hot spots, and thermal expansion.

The uplifts of central Asia (Baikal and Tien Shan) are a SW-NE striking arch [Zorin and Florensov, 1979]. Rifts are associated with these uplifts, confined to their axes. Normal faults and earthquake local mechanisms suggest that the Baikal Rift is an extensional feature oriented across the arch. The upper mantle beneath the arch has low P_n, high Q, and low density, and it is in approximate isostatic

TABLE 2. Major Plateaus and High Plains of the World (Kossinna, 1933)

	Mean Elevation, m	Area, 1000 km^2
German Subalpine Foreland	500	35
Iceland Plateau	600	70
New Castille Plateau	600	60
Old Castille Plateau	700	70
French Massif Central	700	70
Scandinavian Highlands	700	350
Deccan Plateau	800	400
Shotts (Atlas) Plateau	800	80
Nejd Plateau (Arabia)	900	700
Anatolian Plateau (Asia Minor)	1,000	500
Kalahari	1,000	2,100
Tarim Basin	1,100	600
Gobi	1,100	1,650
East African Lakes Plateau	1,200	1,000
Iranian Plateau	1,300	1,000
Great Basin (United States)	1,500	600
Colorado Plateau	1,800	500
Greenland Ice Plateau	1,900	1,870
Armenian Highlands	2,000	300
Yunnan Highlands	2,000	300
Mexican Plateau (Altiplano)	2,000	350
Ethiopian Plateau	2,200	450
Antarctic Ice Plateau	2,500	12,800
Ecuador Plateau (Altiplano)	3,000	15
Bolivian Plateau (Altiplano)	3,800	350
Pamir Plateau	4,000	100
Tibetan Plateau	4,500	2,000
Tharsis area on Mars	10,000	16,000

Fig. 1. Map showing the location of the Colorado Plateau and the surrounding area.

balance. It is inferred that the asthenosphere extends to the base of the crust under the uplift. Lateral difference in lithosphere thickness alone, however, does not appear to account for the data, and laterally heterogeneous asthenospheric properties are implied, namely, a density contrast of 0.005 g/cc. Zorin and Florensov [1979] conclude that SW-NE compression (from Tibet) cannot generate all the structure, but that uprise of hot mantle is required.

The Tharsis region on Mars is a large (16 million km^2) area in which the structural uplift is believed to be about 10 km. A regional free air gravity anomaly of 500 mgal is associated with Tharsis, flanked by gravity lows of 200 mgals. The Bouguer anomaly of minus 700 mgal exists over Tharsis. Its surface is covered by apparently young volcanoes and volcanic plains. Sleep and Phillips [1979] have proposed that the uplift is supported by a low density root extending several hundred km into the mantle (a Pratt isostatic model). Phillips [1978] described a number of possible evolutionary scenarios, but the important point is that plateau uplifts do occur on planets without well-developed plate tectonics.

Colorado Plateau

The Colorado Plateau occupies most of eastern Utah and parts of Colorado, Arizona, and New Mexico (Figure 1). Prominent among reviews of the geology or tectonics of the region are those by Hess [1954], Hunt [1956], Kelley [1955],

Gillully [1963], Eardley [1962], Pakiser [1963], and the plate tectonic models of Atwater [1970], Lipman et al. [1972], Christiansen and Lipman [1972], and Coney [1972]. Most rocks now exposed are undeformed Mesozoic sediments at an average elevation of about 2 km above sea level. Tertiary igneous rocks that include basaltic to silicic volcanics, kimberlite and lamprophyre dikes, and intermediate composition intrusive laccolithic centers are exposed within the province [Williams, 1936; Wenrich-Verbeek, 1979]. Basaltic volcanism predominates and is associated with tensional structures around the plateau margins; the source of the basalts is mantle peridotites [Moore, 1978]. Xenoliths and geophysical data together provide fairly specific constraints on the nature of the lower crust and upper mantle; the upper mantle is apparently hot, possibly even partially molten.

The lower crust and upper mantle xenoliths found in volcanic rocks show hydration effects [Smith, 1978; Smith and Levy, 1976] not observed around the plateau margins [Padovani, 1977; Padovani et al., 1978]. Xenoliths suggest that the lower crust consists of high-rank mafic metaigneous rocks and the upper mantle of spinel peridotite at shallow depths and garnet peridotite at greater depth [Kay and Kay, 1978; McGetchin and Silver, 1972] (Figure 2). An assortment of apparently deep-seated eclogitic rocks are present, exhibiting low-temperature, moderate-pressure mineralogy but with both prograde and retrograde textures [Helmstaedt and Schulze, 1977; 1978], which may be fragments of a subducted plate.

New surface wave data [Keller et al., 1979] indicate that the crust under the Colorado Plateau is 45 km thick and is shieldlike, whereas the normal faulted regions surrounding the plateau have thinner crusts. A very low P_n velocity (7.5 km/s) within the transition zone was indicated by new refraction data. The upper mantle p wave velocity under the plateau is anomalously low, 7.8 km/s. A refraction line near Socorro in the Rio Grande Rift, just off the southeast margin of the plateau, revealed a crustal thickness of 33 km, an apparent P_n of 7.6 km/s, and a strong intracrustal reflection where a magma chamber had been postulated independently by Sanford and by deep seismic reflection (COCORP). (The Consortium for Continental Reflection Profiling (COCORP) uses state-of-the-art multichannel seismic reflection methods to study a variety of geological problems in both the eastern and western U.S.) New, deep bore-hole heat flow measurements [Reiter et al., 1979] reaffirm that the surface flux in the plateau is about 70 mWm^{-2}, or 1.6 HFU ($\mu cal\ cm\ s^{-1}$). Local exceptions seem to be due to hydrothermal circulation and recent volcanism. This flux is higher than the average for areas of Precambrian crust and suggests that the crust and upper mantle beneath the plateau are unusually hot. As Thompson and Zoback [1979] have shown, magnetic

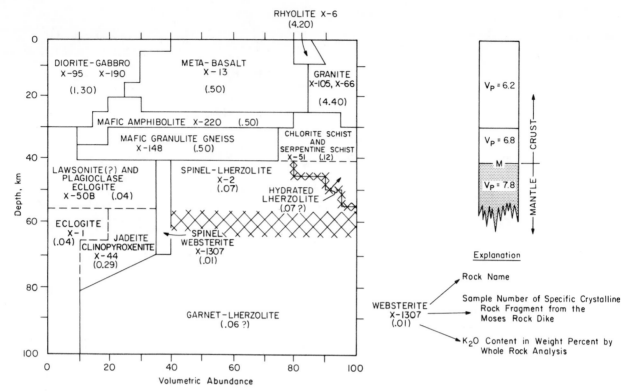

Fig. 2a. Estimated volumetric abundance of rocks constituting the crust and upper mantle under the Moses rock dike. Sample numbers refer to specific rocks described by McGetchin [1968]. (Reprinted from McGetchin and Silver, 1972.)

anomalies indicate a more conductive, presumably hotter, upper mantle beneath the plateau. The expansion caused by this heating may support the elevation of the plateau. The narrow zones of even thinner lithosphere which surround the plateau may supply the force necessary to cause the observed E-W compressive stress in the plateau and to keep the plateau from breaking appart. Recent work by Chapman et al., [1978] places geophysical constraints on the nature of the boundary between the Colorado Plateau and Basin-Range (see Figure 3), namely, abrupt changes in seismic energy release, heat flow, crustal thickness, and P velocity, but absence of significant elevation difference.

Cenozoic gravels and mid-Tertiary volcanics are crucial for interpreting the specifics of the uplift -- its timing and location [Otton and Brooks, 1978; Young, 1979]. In fact, there is some disagreement about whether there is significant relative uplift between the Colorado Plateau and Basin-Range provinces, both of which are structurally high [Damon, 1979; Peirce et al., 1979]. There is no doubt that the entire region (western United States) is structurally, as well as topographically, elevated in relation to sea level or the midcontinent. A structure contour map drawn on the top of the Precambrian would show that the mountains of the Basin-Range stand

even higher than the Colorado Plateau. Hence while the Colorado Plateau is a rather dramatic topographic high, its structural elevation relative to the Basin-Range province on the west, south, and east (Rio Grande Rift) is lower [Kelley, 1979; Mayer, 1979]. The current view [Lucchitta, 1979; Shoemaker, 1978; Young, 1979] is that prior to about 24 Ma ago the Colorado Plateau was topographically low, had low relief and internal drainage, and had not yet developed an integrated through-flowing river system (the Colorado) which exited to the sea. There was significant faulting (including the Basin-Range episode) in the Miocene between 18 and 20 Ma ago, following which time the drainage became integrated into the ancestral Colorado River system (Pliocene time), which flowed off the plateau to the west. The Peach Spring tuff is a key marker unit in pinning down these relations. It appears that prior to eruption of this ash flow, the Basin-Range province stood topographically high relative to the Colorado Plateau; after about 10 Ma the reverse was true. The bulk of the regional uplift, some 1 km, occurred in the Miocene during this 8 Ma span (between about 18 and 10 Ma ago), and the Colorado Plateau province has stood high since. The regional uplift is obscured by contemporaneous Basin-Range faulting and relative subsidence west of the plateau.

CRUSTAL MODELS

ROLLER (1965)			PRODEHL (1970)			PADOVANI et al. (1978)
Depth km	Velocity km/s	Comment	Depth km	Velocity km/s	Comment	Rock Types
0			0			
	6.2	constant velocity		6.1 - 6.2	constant velocity	granite
25		discontinuity	7	6.1 - 6.2	mid-crustal gradient zone	mixed metamorphic and igneous
	6.8		33	6.7 - 6.8		
		constant velocity			high gradient zone	mafic gneiss
			37			
42		Mohorovicic discontinuity	42 - 45	7.6	transitional Mohorovicic discontinuity	
	7.8		47			
				7.8		

Fig. 2b. Crustal models for the Colorado Plateau based on crustal xenoliths from San Juan County, Utah. (Reprinted from Padovani et al., 1978.)

Stratigraphic and structural relationships are especially well exposed and well studied along the SW margin of the Colorado Plateau; the Peach Spring tuff is a distinctive widespread marker useful in defining both the time of uplift and the position of the plateau boundary with the Basin-Range [Lucchitta, 1979; Young, 1979].

Lucchitta [1979] has pointed out that an important measure of the uplift of the plateau with respect to sea level is provided by upwarping and faulting of sea level or near sea level deposits from the mouth of the Colorado to the mouth of the Grand Canyon. He concludes (1) that before Basin-Range faulting, the plateau was lower structurally and topographically than the country to the west; (2) that at the end of Basin-Range faulting and deposition (about 8 Ma ago) the plateau was 1.1 km higher than the top of the adjacent basin fill, 4.3-5.7 km higher than the floor of adjacent basins, but still lower than nearby ranges; and (3) that since about 5 Ma ago, at least 880 m of uplift of the

plateau and Basin-Range have occurred through upwarping and faulting.

There was a general consensus that the plateau we see in Arizona today has been inherited from the modification of a regional northeast-sloping surface related to a broader uplift of much of the western United States dating from the Laramide. This old surface, extensively eroded in early Tertiary time, is covered by Oligocene volcanics on the western and southern margin of the plateau [Peirce et al., 1979; Young, 1979].

Uplift Mechanisms

More than a dozen different mechanisms have been proposed to account for the plateau uplifts (see Table 3), including: (1) thermal expansion of the lithosphere due to a deep mantle plume or hot spot; (2) thermal expansion due to overriding and subduction of a ridge; (3) thermal expansion due to shear heating, accompanying relative motion along the lithosphere-asthenosphere inter-

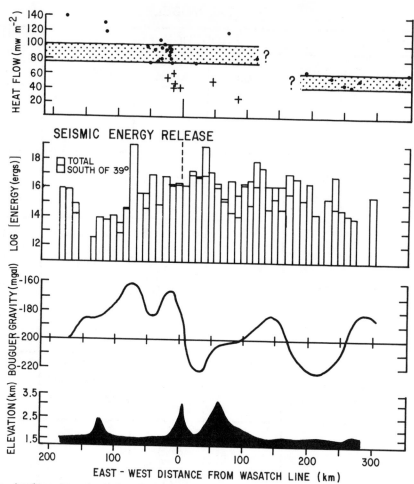

Fig. 3a. Geophysical properties along an east-west cross section through the Colorado Plateau. The crustal thickness beneath the Colorado Plateau is greater than that on either side; its heat flow is above normal; its electrical conductivity is high; and the p wave veolcity is low. (Reprinted from Thompson and Zoback, 1978.)

face; (4) volumetric expansion accompanying partial melting; (5) hydration reactions within the mantle, such as serpentinization; (6) introduction of volatiles from a deep-seated source beneath, due to humite; (7) depletion of 'fertile mantle' in garnet and iron due to partial melting, producing a chemically depleted, refractory residuum of lower density than the original garnet lherzolite; (8) 'crustal thickening' due to a horizontal mass transfer of material (by an unspecified process); (9) underplating or subduction at a very shallow (or even horizontal) angle; (10) metamorphic (solid state) reactions such as basalt-eclogite or spinel-olivine; (11) 'simple' subduction; (12) cessation of subduction and complex reactions accompanying thermal reequilibration of a slab which has stopped; (13) reactivation of preexisting low-angle (listric) thrust faults; and finally (14) detachment and foundering of large pieces of the lithosphere beneath the continents (termed 'delamination')

and flow of mantle material in the asthenophere to replace it. In addition, in classic Chinese and Japanese literature, the burping of a large subterranean frog was believed to be responsible for earthquakes and eruptions. The consensus seems to be that on the continents, uplifts of regional extent are invariably related to plate motions -- subduction or fission. On the ocean floor they are not. On Mars we have insufficient data to discuss the problem thoroughly, tempting though it is to speculate on the possible implications of Tharsis for martian plates in some incipient form.

Conclusions

Several conclusions can be made with respect to plateau uplifts. Upper mantle structure, composition, and thermal processes are the key to understanding the ultimate causes of uplift in virtually every case. It is generally agreed

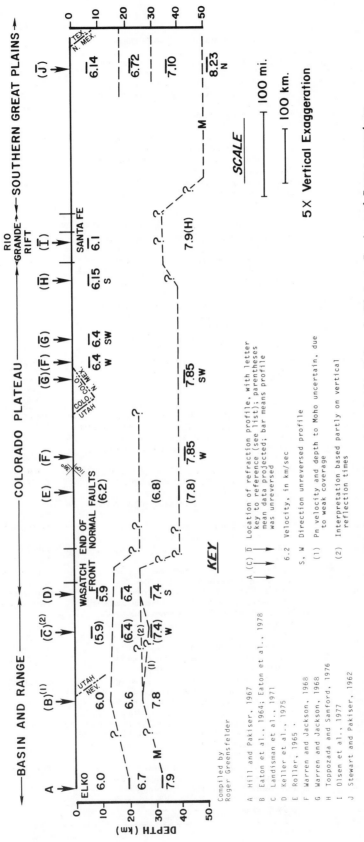

Fig. 3b. Summary of geophysical properties through the Colorado Plateau–Basin and Range province boundary (zero on abscissa). (Reprinted from Chapman et al., 1978.)

Compiled by
Roger Greensfelder

A Hill and Pakiser, 1967
B Eaton et al., 1964; Eaton et al., 1978
C Landisman et al., 1971
D Keller et al., 1975
E Roller, 1965.
F Warren and Jackson, 1968
G Warren and Jackson, 1968
H Toppozada and Sanford, 1976
I Olsen et al., 1977
J Stewart and Pakiser, 1962

KEY

A (C) D̄ → Location of refraction profile, with letter
              key to reference (see list); parentheses
              mean data projected; bar means profile
              was unreversed

6.2 Velocity, in km/sec

S, W Direction unreversed profile

(1) Pn velocity and depth to Moho uncertain, due
     to weak coverage

(2) Interpretation based partly on vertical
     reflection times

SCALE

|———————————| 100 mi.

|———| 100 km.

5 X Vertical Exaggeration

TABLE 3. Summary of Uplift Mechanisms

Mechanism	Description Remarks, and Possible Tectonic Setting	Author or Proponent
1. Thermal expansion due to deep mantle plume or hot spot	Moving lithosphere overriding a hot spot or plume is heated; expansion causes uplift. Lithosphere may thin as well, and volcanism may follow melting. Hawaii is a good example; eastern Australia may be another.	Shaw and Jackson (1973) Morgan (1972) Crough (1979)
2. Thermal expansion due to overriding and subduction of ridge	Subduction of a midocean ridge or hot spreading center would introduce abnormally hot rocks into the upper mantle; expansion resulting could cause uplift and extensive volcanism.	Lipman et al. (1972) DeLong and Fox (1977)
3. Thermal expansion due to shear heating along lithosphere-asthenosphere interface	Viscous heating of shear zone between lithosphere and asthenosphere can produce expansion and resulting uplift.	Melosh and Ebel (1979) Shaw and Jackson (1973)
4. Expansion accompanying partial melting	The volume of fusion for basaltic magma is about 8%; partial melting of the upper mantle would produce expansion.	
5. Hydration reactions such as serpentinization	Volatiles, introduced into a cool upper mantle could produce serpentine, with about 10% volumetric expansion. Hess suggested this for the Colorado Plateau in 1954.	Hess (1954)
6. Introduction of volatiles due to deep-seated dehydration of hydrous minerals	Humite, amphibole carbonate, and phlogopite are observed in deep-seated nodules in kimberlite; dehydration of these minerals and upward transport of the volatiles would contribute to or cause expansion and possibly partial melting.	McGetchin and Silver (1972)
7. Expansion due to depletion of 'fertile' mantle in garnet and iron resulting from basalt genesis	Subcontinental and suboceanic lithospheres (upper mantles) have similar densities and seismic velocities but very different temperatures.	Jordan (1975)
8. Crustal thickening due to horizontal transfer of mass in the lower crust	Process described by Gillully prior to wide acceptance of plate tectonics to explain apparent mass deficiency under the Basin-Range and coincident elevation of the Colorado Plateau; details of process unspecified. Hess proposed crustal thickening by compression.	Gillully (1963) Hess (1962)
9. Underplating or subduction at very shallow angle	Metamorphic assemblages, such as high-pressure low-temperature eclogites suggest that ophiolite-like plates may exist beneath the Colorado Plateau; some believe that thrust angles could be nearly horizontal.	Helmstaedt and Schulze (1977)
10. Metamorphic (solid state) reactions as eclogite-basalt or spinel-olivine	Deep-seated solid-state reactions; would require heating, in general, to drive reactions to expansions.	O'Connell and Wasserburg (1967) Lovering (1958) Green and Ringwood (1967)
11. "Simple" subduction and/or continental collision during subduction	High plateaus of South America and Tibet are examples of simple subduction processes. The Colorado Plateau may have been underlain by a subduction zone prior to 30 Ma, when the East Pacific Rise intersected the North American plate.	Lipman et al. (1972) Coney (1972) Damon (1979) Atwater, (1970)
12. Cessation of subduction and resulting thermal equilibration of static slab	When subduction stops, the cold downgoing slab heats up. The resulting thermal equilibration and metamorphic reactions (both retrograde and prograde) are complex but will produce expansion, metamorphism, and possibly volcanism.	Christiansen and Lipman (1972) Thompson and Zoback (1979) Silver and McGetchin (1978)
13. Isolation of Colorado Plateau by listric normal faulting in surrounding areas	The Colorado Plateau is part of a much broader Rockly Mountain uplift. West, south, and east of the Colorado Plateau, crustal attenuation is achieved by stretching that is accompanied by listric normal faulting. In this fashion the Colorado Plateau becomes isolated and remains intact. The listric normal faults are preferentially located along preexisting earlier thrust faults.	Bally (1978)
14. Lithospheric delamination, i.e., detachment and foundering of large pieces of subcontinental lithosphere	Owing to cooling, lithosphere detaches from crust and sinks into warmer, more viscous, and denser asthenosphere. Counterflow of asthenosphere warms crustal rocks above; result is progressive uplift and volcanism.	Bird (1979)

that the Colorado Plateau could not be separated from the regional uplift of western North America during Tertiary time. The uplift of the Colorado Plateau province and adjacent regions can be bracketed in time between about 10 and 18 Ma, and the magnitude of upward movement was about 2 km. Available data (P, S velocity structure, Q, electrical resistivity profiles, heat flow, and petrology of volcanics and xenoliths) strongly suggest partial melting within the upper mantle under much of the western United States. The 'Colorado Plateau uplift problem' then becomes one of explaining (1) the ultimate cause of the regional uplift of the American West and (2) why the Colorado Plateau survived as a structural and topographic entity while the country surrounding it was being severely modified by faulting and extensive erosion. It is apparent that the structural and topographic unit known as the Colorado Plateau is being 'nibbled' away from all sides by active erosion and faulting.

It was pointed out, partly in jest, by George Thompson that Dutton in 1892 suggested that uplifts are due primarily to expansion. No less that 14 fundamental uplift mechanisms can be defined (see Table 3), essentially all involving subduction, plate motion, and/or expansion due to heating or partial melting, but there is a clear need for new data. Comparisons among plateaus on continents, on the sea floor, and even on other planets (the Tharsis uplift on Mars) will provide an important overview, but it now appears that uplifts are most commonly associated either (1) with subduction or its direct effects or (2) with deep-seated thermal disturbances which result in expansion and uplift.

It is important to conclude with the question: What new data do we need to address crucial questions for the Colorado Plateau (and other plateau uplifts of the world)? The answers will come from many disciplines but the most important new data are likely to be contributed by (1) geologic and geomorphologic mapping of relations on the margins of uplifts, especially supported by absolute age determinations of associated volcanics which will constrain the amount and timing of uplifts; (2) petrology of volcanic rocks as clues to their origin and relationship to subducting plate margins; (3) petrology of xenoliths in volcanic rocks which provide important clues to the composition and state of the lower crust and upper mantle and, in particular, metamorphic events occurring at depth through quantitative constraints on pressure, temperature, and chronology of these rocks (there is a critical need for age-dating methods applicable to mafic and ultramafic lower crustal and upper mantle xenoliths); and (4) geophysics which will continue to provide the bulk of data for the physical properties of the lower crust and the upper mantle. New programs such as Continental Drilling, COCORP, satellite geodesy (e.g., NASA's proposed geodynamics program), and consortia approaches to topical studies may help considerably in providing new data.

Like the Colorado Plateau itself, the problem of plateau uplifts is being nibbled at from all sides -- progress is steady, if not spectacular.

Acknowledgments. This work was performed at the Lunar and Planetary Institute, which is operated by the Universities Space Research Association, under contracts NSR 09-051-001 and NASW-3389 with the National Aeronautics and Space Administration. Helpful reviews of an earlier manuscript were provided by Bert Bally, Ivo Lucchitta, and Russell Merrill. This is Lunar and Planetary Institute Contribution 409.

References

Atwater, T., Implications of plate tectonics for the Cenozoic tectonic evolution of western North America, Bull. Geol. Soc. Am., 81, 3513-3536, 1970.

Bally, A. W., The global view of plateaus, Conference on Plateau Uplift: Mode and Mechanism, Flagstaff, Arizona, 14-16 August 1978 (oral presentation), 1978.

Bird, P., Continental delamination and the Colorado Plateau, J. Geophys. Res., in press, 1980.

Bond, G. C., Relative uplifts of large continental areas (Abstract), in Papers Presented to the Conference on Plateau Uplift: Mode and Mechanism, pp. 1-3, Lunar and Planetary Institute, Houston, 1978.

Burke, K., Categories of plateaus on Earth (Abstract), in Papers Presented to the Conference on Plateau Uplift: Mode and Mechanism, pp. 8-9, Lunar and Planetary Institute, Houston, 1978.

Chapman, D. S., K. P. Furlong, R. B. Smith, and D. J. Wechsler, Geophysical characteristics of the Colorado Plateau and its transition to the Basin and Range Province in Utah (Abstract), in Papers Presented to the Conference on Plateau Uplift: Mode and Mechanism, pp. 10-12, Lunar and Planetary Institute, Houston, 1978.

Christiansen, R. L. and P. W. Lipman, Cenozoic volcanism and plate-tectonic evolution of the western United States 2. Late Cenozoic, Phil. Trans. Roy. Soc. London, Ser. A, 271, 249-284, 1972.

Coney, P. J., Cordilleran tectonics and North America plate motion, Am. J. Sci., 272, 603-628, 1972.

Crough, S. T., Thermal origin of the plateaus surrounding mid-plate, hot-spot volcanoes (Abstract), in Papers Presented to the Conference on Plateau Uplift: Mode and Mechanism, p. 13, Lunar and Planetary Institute, Houston, 1978.

Crough, S. T., Hotspot epeirogeny (in Proceedings of the Conference on Plateau Uplift: Mode and Mechanism, Flagstaff, Arizona, 14-16 August 1978), Tectonophysics, 61, 321-333., 1979.

Damon, P. E., Continental uplift at convergent boundaries (in Proceedings of the Conference on Plateau Uplift: Mode and Mechanism, Flagstaff,

Arizona, 14-16 August 1978), Tectonophysics, 61, 307-319, 1979.

De Long, S. E. and P. J. Fox, Geological consequences of ridge subduction, in Maurice Ewing Series, vol. 1, Island Arcs, Deep Sea Trenches and Back-Arc Basins, pp. 221-228, 1977.

Dutton, C. E., On some of the greater problems of physical geology, Phil. Soc. of Washington, 11, 51-64, 1892.

Eardley, A. J., Structural Geology of North America, 2nd ed., Harper and Row, New York, 743p., 1962.

Eaton, J. P., J. Healy, W. D. Jackson, and L. C. Pakiser, Upper mantle velocity and crustal structure in the eastern Basin and Range province determined from SHOAL and chemical explosions near Delta, Utah (Abstract), Bull. Seism. Soc. Am., 54, 1567, 1964.

Eaton, G. P., D. R. Mabey, R. R. Wahl, and M. D. Kleinkopf, Regional gravity and tectonic patterns: their relation to Cenozoic epeirogeny and lateral spreading in the western Cordillera, in Cenozoic Tectonics and Regional Geophysics of the Western United States, edited by R. B. Smith and G. P. Eaton, Geol. Soc. Am. Memoir 152, 1978.

Gillully, J., The tectonic evolution of the western United States, Quart. J. Geol. Soc., London, 119, 113-174, 1963.

Green, D. H. and A. E. Ringwood, An experimental investigation of the gabbro to eclogite transformation and its petrological applications, Geochim. Cosmochim. Acta, 31, 767-833, 1967.

Helmstaedt, H. and D. J. Schulze, Type A-Type C eclogite transition in a xenolith from Moses Rock diatreme -- further evidence of metamorphosed ophiolites beneath the Colorado Plateau (Abstract), in 2nd International Kimberlite Conference, Abstract Volume, Santa Fe, N. Mex., 1977.

Helmstaedt, H. and D. J. Schulze, Petrologic constraints for upper mantle models of the Colorado Plateau (Abstract), in Papers Presented to the Conference on Plateau Uplift: Mode and Mechanism, Lunar and Planetary Institute, Houston, pp. 16-18, 1978.

Hess, H. H., Serpentines, orogeny and epeirogeny, in Crust of the Earth -- a symposium, edited by A. Poldervaart, Geol. Soc. Am. Spec. Pap. No. 62, pp. 391-408, 1954.

Hess, H. H., History of ocean basins, in Petrologic Studies: a Volume in Honor of A. F. Buddington, edited by A. E. J. Engle, pp. 599-620, 1962.

Hill, D. P. and L. C. Pakiser, Seismic-refraction study of crustal structure between the Nevada test site and Boise, Idaho, Geol. Soc. Am. Bull., 78, 685-704, 1967.

Holmes, A., Principles of Physical Geology, The Ronald Press, New York, 1288p., 1964.

Hunt, C. B., Cenozoic geology of the Colorado Plateau, U. S. Geol. Surv. Prof. Pap. No. 279, 99p., 1956.

Illies, J. H., C. Prodehl, H. -U. Schmincke, and

A. Semmel, The Quaternary uplift of the Rhenish Shield in Germany, (in Proceedings of the Conference on Plateau Uplift: Mode and Mechanism, Flagstaff, Arizona, 14-16 August 1978) Tectonophysics, 61, 197-225, 1979.

Jordan, T. H., Laterial heterogeneity and mantle dynamics, Nature, 257, 745-750, 1975.

Kailasam, L. N., Plateau uplift in peninsular India (Abstract), in Papers Presented to the Conference on Plateau Uplift: Mode and Mechanism, Lunar and Planetary Institute, Houston, pp. 22-24, 1978.

Kay, R. and S. M. Kay, Regional variations of the lower continental crust: inferences from magmas and xenoliths (Abstract), in Papers Presented to the Conference on Plateau Uplift: Mode and Mechanism, Lunar and Planetary Institute, Houston, p. 25, 1978.

Keller, G. R., L. W. Braile, and P. Morgan, Crustal structure, geophysical models and contemporary tectonism of the Colorado Plateau, (in Proceedings of the Conference on Plateau Uplift: Mode and Mechanism, Flagstaff, Arizona, 14-16 August 1978) Tectonophysics, 61, 131-147, 1979.

Keller, G. R., R. B. Smith, and L. W. Braile, Crustal structure along the Great Basin - Colorado Plateau transition from seismic refraction measurements, J. Geophys. Res., 80, 1093-1098, 1975.

Kelley, V. C., Regional tectonics of the Colorado Plateau and relationship to origin and distribution of uranium, Univ. New Mex. Publ. Geol. 5, 120p., 1955.

Kelley, V. C., Tectonics of the Colorado Pateau and new interpretation of its eastern boundary (in Proceedings of the Conference on Plateau Uplift: Mode and Mechanism, Flagstaff, Arizona, 14-16 August 1978) Tectonophysics, 61, 97-102, 1979.

Kossinna, E., Die Erdoberflache, in Handbuch d. Geophysik, edited by B. Gutenberg, Borntraeger, Berlin, vol. 2, pp. 869-954, 1933.

Landisman, M., S. Mueller, and B. J. Mitchell, Review of evidence for velocity inversions in the continental crust, in The Structure and Physical Properties of the Earth's Crust, edited by J. G. Heacock, Am. Geophys. Un. Mono. 14, p. 11, 1971.

Lipman, P. W., H. J. Prostka, and R. L. Christiansen, Cenozoic volcanism and plate-tectonic evolution of the western United States, I. Early and Middle Cenozoic, Phil. Trans. Roy. Soc. London, Ser. A., 271, 217-248, 1972.

Lovering, J. F., The nature of the Mohorovicic discontinuity, Trans. Am. Geophys. Union, 39, 947-955, 1958.

Lucchitta, I., Late Cenozoic uplift of the southwestern Colorado Plateau and adjacent lower Colorado River region, (in Proceedings of the Conference on Plateau Uplift: Mode and Mechanism, Flagstaff, Arizona, 14-16 August 1978) Tectonophysics, 61, 63-95, 1979.

Mayer, L., Evolution of the Mogollon Rim in central Arizona, (in Proceedings of the Conference on Plateau Uplift: Mode and Mechanism, Flagstaff, Arizona, 14-16 August 1978) Tectonophysics, 61, 49-62, 1979.

McGetchin, T. R., The Moses Rock Dike: Geology, petrology and mode of emplacement of kimberlite-bearing breccia dike, San Juan County, Utah, Ph. D. Thesis, California Institute of Technology, 405p., 1968.

McGetchin, T. R. and L. T. Silver, A crustal-upper mantle model for the Colorado Plateau based on observations of crystalline rock fragments in the Moses Rock dike, J. Geophys. Res., 77, 7022-7037, 1972.

Melosh, H. J. and J. Ebel, A simple model for thermal instability in the asthenosphere, Geophys. J. Roy. Ast. Sco., in press, 1979.

Moore, R. B., Cenozoic igneous rocks of the Colorado Plateau (Abstract), in Papers Presented to the Conference on Plateau Uplift: Mode and Mechanism, Lunar and Planetary Institute, Houston, pp. 28-30, 1978.

Morgan, W. J., Convection plumes and plate tectonics, Am. Assoc. Petrol. Geol. Bull., 56, 203-213, 1972.

O'Connell, R. J. and G. J. Wasserburg, Dynamics of the motion of a phase change boundary to changes in pressure, Rev. Geophys., 5, 329-410, 1967.

Olsen, K. H., G. R. Keller, J. N. Stewart, E. F. Homuth, and D. J. Cash, A seismic refraction study of crustal structure in the Rio Grande Rift, New Mexico, Trans. Am. Geophys. Union, 58, 1184, 1977.

Otton, J. K. and W. E. Brooks, Jr., Tectonic history of the Colorado Plateau margin, Date Creek Basin and adjacent areas, west-central Arizona (Abstract), in Papers Presented to the Conference on Plateau Uplift: Mode and Mechanism, Lunar and Planetary Institute, Houston, p. 31-33, 1978.

Padovani, E. R., Granulite facies xenoliths from Kilborne Hole maar and their bearing on deep crustal evolution, Ph.D. thesis, U. Texas, Dallas, 1977.

Padovani, E. R., J. Hall, and G. Simmons, V_p measurements on crustal xenoliths from San Juan County, Utah (Abstract), in Papers Presented to the Conference on Plateau Uplift: Mode and Mechanism, Lunar and Planetary Institute, Houston, p. 34-35, 1978.

Pakiser, L. C., Structure of the crust and upper mantle in the western United States, J. Geophys. Res. 68, 5747-5756, 1963.

Peirce, H. W., P. E. Damon, and M. Shafiqullah, An Oligocene (?) Colorado Plateau edge in Arizona, (in Proceedings of the Conference on Plateau Uplift: Mode and Mechanism, Flagstaff, Arizona, 14-16 August 1978) Tectonophysics, 61, 1-24, 1979.

Phillips, R. J., Mars: the Tharsis uplift (Abstract), in Papers Presented to the Conference on Plateau Uplift: Mode and Mechanism,

Lunar and Planetary Institute, Houston, p. 40-42, 1978.

Prodehl, C., Seismic refraction study of crustal structure in the western United States, Bull. Geol. Soc. Am., 81, 2629-2646, 1970.

Reiter, M., A. J. Mansure, and C. Shearer, Geothermal characteristics of the Colorado Plateau, (in Proceedings of the Conference on Plateau Uplift: Mode and Mechanism, Flagstaff, Arizona, 14-16 August 1978) Tectonophysics, 61, 183-195, 1979.

Roller, J. C., Crustal structure in the eastern Colorado Plateau province from seismic refraction measurements, Bull. Seism. Soc. Am., 55, 107-119, 1965.

Sclater, J. G., R. N. Anderson, and M. L. Bell, The elevation of ridges and the evolution of the central eastern Pacific, J. Geophys. Res., 76, 7888-7915, 1971.

Sclater, J. G. and J. Francheteau, The implications of terrestrial heat flow observations on current tectonic and geochemical models of the crust and upper mantle of the Earth, Geophys. J. Roy. Ast. Soc., 20, 509-542, 1970.

Shaw, H. R. and E. D. Jackson, Linear island chains in the Pacific: Result of thermal plumes or gravitational anchors, J. Geophys. Res., 78, 8634-8652, 1973.

Shoemaker, E. M., The Colorado Plateau problem, Conference on Plateau Uplift: Mode and Mechanism, Flagstaff, Arizona, 14-16 August 1978 (oral presentation), 1978.

Silver, L. T. and T. R. McGetchin, The nature of the basement beneath the Colorado Plateau and some implications for plateau uplifts (Abstract), in Papers Presented to the Conference on Plateau Uplift: Mode and Mechanism, Lunar and Planetary Institute, Houston, p. 45-46, 1978.

Sleep, N. H. and R. J. Phillips, An isostatic model for the Tharsis province, Mars, J. Geophys. Res., 6, 803-806.

Smith, D., Eclogite and hydrated peridotite inclusions in volcanic rocks on the Colorado Plateau (abstract), in Papers Presented to the Conference on Plateau Uplift: Mode and Mechanism, Lunar and Planetary Institute, Houston, p. 47-48, 1978.

Smith, D. and S. Levy, Petrology of the Green Knobs diatreme and implications for the upper mantle below the Colorado Plateau, Earth and Planet. Sci. Lett., 29, 107-125, 1976.

Stewart, S. W. and L. C. Pakiser, Crustal structure in eastern New Mexico interpreted from GNOME explosion, Bull. Seism. Soc. Am, 52, 1017-1020, 1962.

Theilen, Fr. and R. Meissner, A comparison of crustal and upper mantle features in Fennoscandia and the Rhenish Shield, two areas of recent uplift, (in Proceedings of the Conference on Plateau Uplift: Mode and Mechanism, Flagstaff, Arizona, 14-16 August 1978) Tectonophysics, 61, 227-242, 1979.

Thompson, G. A. and M. L. Zoback, Regional

geophysics of the Colorado Plateau, (in Proceedings of the Conference on Plateau Uplift: Mode and Mechanism, Flagstaff, Arizona, 14-16 August 1978) Tectonophysics, 61, 149-181, 1979.

Toppozada, T. R. and A. R. Sanford, Crustal structure in New Mexico interpreted from the Gasbuggy explosion, Bull. Seism. Soc. Am., 66, 877-886, 1976.

Urrutia-Fucugauchi, J., Lithospheric and crustal evolution of central Mexico (abstract), in Papers Presented to the Conference on Plateau Uplift: Mode and Mechanism, Lunar and Planetary Institute, Houston, p.14-15, 1978.

Warren, D. H. and W. H. Jackson, Surface seismic measurements of the project Gasbuggy explosions at intermediate distance ranges, U. S. Geol. surv. Open File Report 45, 1968.

Wenrich-Verbeek, K. J., The petrogenesis and trace-element geochemistry of intermediate lavas from Humphreys Peak, San Francisco Volcanic Field, Arizona, (in Proceedings of the Conference on Plateau Uplift: Mode and Mechanism, Flagstaff, Arizona, 14-16 August 1978) Tectonophysics, 61, 103-129, 1979.

Williams, H., Pliocene volcanos of the Navajo-Hopi country, Bull. Geol. Soc. Am., 47, 111-172, 1936.

Young, R. A., Laramide deformation, erosion, and plutonism along the southwestern margin of the Colorado Plateau, (in Proceedings of the Conference on Plateau Uplift: Mode and Mechanism, Flagstaff, Arizona, 14-16 August 1978) Tectonophysics, 61, 25-47, 1979.

Zorin, Yu. A. and N. A. Florensov, On geodynamics of Cenozoic uplifts in central Asia, (in Proceedings of the Conference on Plateau Uplift: Mode and Mechanism, Flagstaff, Arizona, 14-16 August 1978) Tectonophysics, 61, 271-283, 1979.

MODELS OF GLACIAL ISOSTASY AND RELATIVE SEA LEVEL

W.R. Peltier*

Department of Physics, University of Toronto, Toronto, Ontario M5S 1A7

Abstract. The interpretation of relative sea level data from the Quaternary period demands a global model of the phenomenon of glacial isostatic adjustment. The main parameter of such a model is the effective viscosity of the planetary mantle and the procedure for determining this parameter from the observations is a problem in the theory of inference. The purpose of this article is to provide a non-mathematical description of the main physical ingredients that an appropriate model must include if it is to prove an effective vehicle for the interpretation of the global data set. An attempt is made to provide an historical perspective of the way in which such models have evolved, particularly during the past decade, and to trace the simultaneous revisions of our view of the effective viscosity of the interior which has accompanied the evolution of theory.

Introduction

Twenty thousand years ago, during the maximum of the Wurm-Wisconsin glaciation, a large fraction of the northern hemisphere continental surface was covered by thick ice sheets which have since disintegrated. The mass contained in these ice sheets was on the order of 10^{23} grams or roughly one part in 10^6 of the total mass of the planet. When these ice sheets melted, the addition of meltwater to the ocean basins led to an increase of approximately 80 metres in mean global sea level. This deglaciation event (which began ca. 18ka BP) involved two major ice sheets: the Laurentide, which covered all of Canada and parts of the northern United States, and the Fennoscandian ice sheet, which was centred over Sweden. Neither of the presently existing ice caps over Antarctica and Greenland were significantly affected by the climatic change, a fact which remains to a large extent enigmatic. Even the physical processes that were responsible for the climatic change are themselves the subject of current debate. In spite of the fact that we do not yet understand in detail why the ice sheets re-

treated, we can nevertheless learn a great deal about the planetary interior from a study of its response to this cataclysmic redistribution of surface mass. This is the subject of glacial isostasy.

The best presently available data (e.g. Hays, Imbrie and Shackleton, 1976), from which it is possible to infer the long time scale variations in the extent of northern hemisphere ice coverage, suggest that prior to deglaciation the major ice sheets had existed near their maximum extent for times on the order of $2-3 \times 10^4$ years. If this inference is correct, then we may argue that at glacial maximum (ca. 2×10^4ka BP) the system consisting of the aquasphere, cryosphere, and solid earth was near gravitational equilibrium. When this equilibrium (or quasi-equilibrium) was upset by massive deglaciation, the resulting unbalanced gravitational forces deformed the shape of the planet and redistributed water among and within the ocean basins. The global history of this deformation has been conveniently recorded for our inspection in the relative sea level record at a large number of sites on the Earth's surface. The time scale spanned by these data (ca. 0–20ka BP) is such that radiocarbon dating provides a fairly reliable technique for determining the age of relict beaches found in most locations. Given the age of each of a sequence of raised or submerged beaches, and direct measurements of their "vertical" separation from present-day sea level, one may reconstruct a local history of relative sea level. As an example of the kind of information generated in this way we consider the staircase of emerged strandlines shown in Figure 1 (after Hilaire-Marcel and Fairbridge, 1978) which are located in the Richmond Gulf of Hudson Bay near what was the centre of the Laurentide ice sheet. Application of the radiocarbon technique to shells contained within the various beach horizons (ages corrected to give proper sidereal age) leads to the relative sea level curve shown in Figure 2, where I have plotted the height above present sea level of each of the main horizons in the staircase

---

* Alfred P. Sloan Foundation Fellow

Figure 1: Staircase of emerged strandlines in the Richmond Gulf on the southeast coast of Hudson Bay near the center of rebound (after Hilaire-Marcel and Fairbridge, 1978).

against age. Data such as these constitute perhaps the most important observables of glacial isostasy.

From the time derivative of such data we may deduce the present-day rate of emergence locally and the global variation of the sign of this "signature" of the adjustment process is shown in Figure 3 (after Walcott, 1972b). This clearly indicates that regions which were once under the major ice sheets are currently emerging, whereas regions peripheral to them are submerging. This is a pattern which any model of rebound must be able to reproduce. On the basis of the uplift and submergence data we must immediately conclude that no Hookean elastic model of the interior could ever reproduce them. Deglaciation of the surface was essentially complete 6×10^3 years before the present, yet the shape of the planet continued to deform subsequently and this process is not yet complete. If the Earth were Hookean elastic then deformation would have been complete "immediately" upon the cessation of melting.

The geodynamic importance of the phenomenon of glacial isostatic adjustment lies in the fact that from observations like those in Figure 2, and on the basis of an appropriate model, one may deduce the effective viscosity of the planetary mantle. It was because of this utility of the rebound data that Reginald Daly was led to refer to the last major deglaciation as "nature's great experiment". Not only did Nature perform the experiment but she also "remembers" the results. The effective viscosity of the mantle is clearly an important geophysical variable. One example of this importance will suffice to make the point. In the convection hypothesis of continental drift

and sea floor spreading the viscosity determines the rate at which mantle material circulates in resonse to a radial temperature gradient which is superadiabatic. It thus (perhaps indirectly) controls the surface plate velocities. Since convection transports heat and since the efficiency of convective heat transport depends upon the circulation time, the viscosity also exerts an important control of the Earth's thermal history. It may be the most important variable. We must therefore enquire as to whether the viscosity of the mantle which one deduces from analysis of the rebound data has the same magnitude as that which is required in the convection hypothesis. Since this is an important question it requires an answer that is as free as possible from ambiguity or at least one in which the ambiguity is carefully quantified. The answer will of course be believable only to the extent that the model employed to obtain it is believable, the two being extricably connected.

It is perhaps not surprising, given the form of rebound data such as those shown in Figure 2, that the first models employed for the purpose of inferring mantle viscosity from them were based upon the assumption that for deformations of such long time scale the Earth behaves as a simple, incompressible, Newtonian viscous fluid (Haskell, 1935,1936,1937; Vening-Meinesz, 1937; Heiskanen and Vening-Meinesz, 1958). The relative sea level data, in the central region of uplift, have the form of a simple expon-

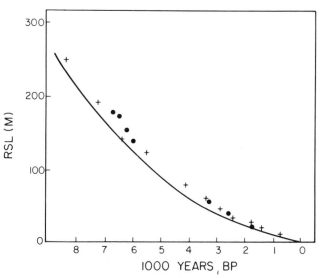

Figure 2: Relative sea level data (solid circles and crosses) from the Richmond Gulf (after Hilaire-Marcel and Fairbridge, 1978). The solid line is the theoretical prediction obtained using the model discussed in section 4, for which the mantle viscosity was assumed constant and equal to 10^{22} Poise (c.g.s. units) in magnitude.

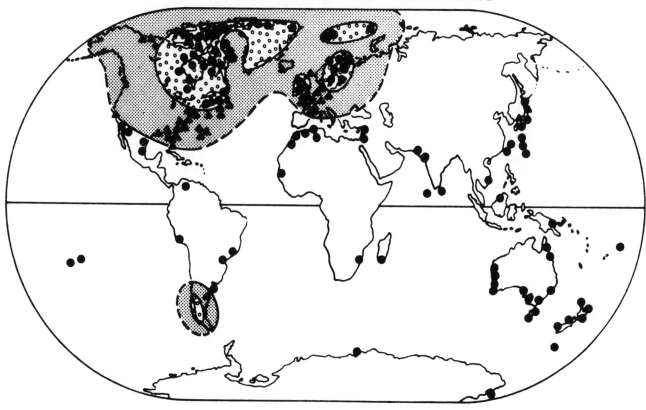

LEGEND

- REGION OF POSTGLACIAL REBOUND
- PERIPHERAL ZONE OF SUBMERGENCE
- DATED MARINE SHELLS ABOVE PRESENT SEA LEVEL
- DATED TERRESTRIAL PEATS BELOW PRESENT SEA LEVEL

Figure 3: Distribution of radiocarbon-dated specimens in the age range 2500-5000 years BP and related to sea level. The solid triangles are peats and other terrestrially deposited materials found below present sea level, and the solid circles are marine shells, corals, or other marine derived materials found above present sea level. Note the zone of submergence surrounding the regions of known rapid postglacial uplift (after Walcott, 1972b).

ential relaxation, and for the Richmond Gulf data the relaxation time is on the order of 1.7×10^3 years. It is perhaps not inappropriate to show immediately how this observed relaxation time may be employed to estimate mantle viscosity. For a uniform viscous sphere of constant density and viscosity the free decay time for a deformation of spherical harmonic degree n is $\tau = (\nu/\rho\ g_s)\ (2n^2 + 4n + 3)/na$ (Peltier, 1974), where ν and ρ are the molecular viscosity and density respectively, g_s is the surface gravitational acceleration and a is the radius of the sphere. Now the scale of the Laurentide ice sheet was such (Peltier and Andrews, 1976)

that the dominant term in its spherical harmonic expansion corresponds to $n \cong 6$. Using $g_s \cong 10^3$ cm s^{-2}, $\rho \cong 3$ gm cm^{-3}, $a \cong 6.371 \times 10^8$ cm and the observed $\tau \cong 1.7 \times 10^3$ years, then $\nu \cong 10^{22}$ Poise. This then is the source of the magic number for the viscosity of the mantle which has existed in the geophysical literature for the past 40 years. It was first deduced, by the authors referenced above, through analysis of the rebound in Fennoscandia, a response which was forced by the melting of an ice sheet of much smaller "horizontal" scale (dominant n = 16) than existed over Canada. The scale is sufficiently small that the models employed were halfspace models rather than spheres, an approximation in which the previous expression relating relaxation time to wavelength reduces to $\tau = 2\nu k_H/g_s$ where k_H is the horizontal wavenumber.

That the viscosity of the mantle suggested by the data from two separate regions in which the uplift is dominated by spatial scales which differ by a factor of three is encouraging. Taken at face value the results suggest (1) that the effective viscosity may not be strongly inhomogeneous laterally, and (2) it does not appear to possess a strong depth dependence. The second of these implications is clearly incorrect since we know that for solids the effective viscosity is a strong function of temperature, regardless whether the creep mechanism is linear or non-linear, and the temperature is a strongly increasing function of depth (at least initially). It has been suggested (Weertman and Weertman, 1975) that this temperature dependence may be represented empirically as $\nu = V \exp(-hT/T_m)$ where h is an empirical constant, T and T_M the temperature and melting temperature respectively, and V is a function of the microphysical properties of the solid and perhaps also of the stress if the creep mechanism is non-linear. Therefore, at least near the surface the viscosity should be considerably in excess of the viscosity at depth. We suspect furthermore that this surface "lithosphere" may be strongly inhomogeneous laterally, reflecting the difference between continental and oceanic regions. The simple homogeneous models are therefore violently at odds with physical intuition and are therefore not "believable". That the viscosity of the earth must be a function of depth can be seen from a second and trivial example. We know that the outer core is liquid and that its viscosity must be enormously less than the viscosity associated with the thermally activated creep of a solid. The mantle is therefore bounded above by a lithosphere of high viscosity and below by a fluid with negligible viscosity in comparison with it. Again, the simple models must be extended if they are not to be in violation of the simplest constraints.

The connection, in the early models of isostatic adjustment, between the spatial scale of a surface deformation and its associated free relaxation time was the main, if not the only, guide employed for the inference of viscosity. In the initial literature of this subject there were two competing interpretations which to a certain extent have both persisted until the present. Van Bemmelen and Berlage (1935), working at the same time as Haskell, supposed that the flow associated with isostatic recovery beneath Fennoscandia was confined to a thin channel extending from the earth's surface to a depth h. Their analysis showed that the relaxation time τ for a deformation of wavenumber k_H was such that $\tau = (12\nu/\rho g\ h^3)$.

$(1/k_H{}^2)$, i.e. that relaxation time decreased as the square of the wavenumber. Remember that the uniform viscosity model predicts that relaxation time is a linearly increasing function of the wavenumber. Assuming $h = 10^2$ km Van Bemmelen and Berlage inferred a viscosity of $\nu = 3 \times 10^{19}$ Poise. Although this idea of thin channel flow near the surface subsequently disappeared from the literature it was resurrected to a certain extent by Takeuchi (1963) and Crittenden (1963). Crittenden in particular compiled data for the relaxation of Pleistocene Lake Bonneville in the Basin and Range Province of the United States and discovered that although the spatial extent of this region differed by an order of magnitude from Fennoscandia, the relaxation times for the two regions differed only slightly, both being on the order of 5000 years. These results seemed to him to support the channel flow model, i.e. the notion that in a thin layer near the surface the viscosity is anomalously low!

McConnell (1968) realized that effective confinement of the flow in a region near the surface could be achieved by assuming that the mantle viscosity was not uniform but increased continuously as a function of depth, a notion which has persisted until recently for reasons which will be discussed below. If one includes in the model the basic structure which *a priori* information demands, then Crittenden's observation is easily reconcilable with uniform mantle viscosity. The effect of a thin lithosphere at the surface of the planet in which the viscosity is very much in excess of that in the underlying mantle is such as to reduce the relaxation time of all harmonics with sufficiently short wavelength. This was pointed out by McConnell (1968) in the context of halfspace models constructed to fit observed relaxation spectra from the Fennoscandian uplift and by Peltier (1980a) for the spherical models necessary to describe the Laurentide data. The latter calculations, which are discussed in Section 2, furthermore show that an Earth model with a viscosity of 10^{22} Poise throughout the mantle and a lithosphere which is about 120 km thick has a relaxation time of approximately 5000 years for both the n = 16 and the n = 160 (dominant Lake Bonneville) harmonics. The Earth model must include a lithosphere if it is to fit the relaxation data. The viscosity of the planet initially decreases with increasing depth in accord with physical intuition. Whether a "low viscosity channel" beneath the lithosphere is required to fit the details of the observed relaxation, it seems to me, is an open question and one which can be resolved only by a detailed "resolving power" calculation. It is nevertheless true that such a region, if it exists at all under the stable continental platforms on which the main ice sheets were located, will influence only the relatively short wavelengths of the relaxation, the same region of the wavenumber spectrum which is governed by the lithosphere.

The lithosphere inhibits the ability of the relaxation to "see" a low viscosity channel. In Cathles' (1975) models of the large scale relaxation, this point is somewhat obscured. The feature of his spherical models which he calls the low viscosity channel extends to the Earth's surface so that these models have the same exterior structure as Crittenden's proposed halfspace models and must be rejected for the same reason. The planet does have a lithosphere and must have if our understanding of the thermally activated creep of solids is not completely erroneous.

In order to estimate the viscosity at increasing depth in the interior we require information on the relaxation times of successively longer horizontal wavelengths in the spectrum. Prior to the late 1960's the largest scale relaxation from which data were available was the Fennoscandia region. In analyzing these data McConnell (1968) was aware, and Parsons (1972) has since formally shown, that the Fennoscandia rebound was sensitive only to viscosity variations in the outer 600 km or so of the planet. In order to extract information about the viscosity from depths in excess of this McConnell (1968) was obliged to invoke extra information. He first assumed that the so-called non-hydrostatic equatorial bulge existed; i.e. that the polar flattening of the sphere (produced by the centrifugal force) was in excess of that which would be in equilibrium with the Earth's current rotation rate (Munk and McDonald, 1960). He further assumed that the excess bulge was produced by glaciation. It then followed from the assumption of its existence that it must have been relaxing very slowly since deglaciation and therefore that the relaxation time of the n = 2 harmonic was in excess of $\tau = 0(10^4)$ years. This requires a high lower mantle viscosity and McConnell described models in which the lower mantle had a viscosity in excess of $\nu = 10^{24}$ Poise. McKenzie (1966, 1967, 1968) came to a similar conclusion - again based upon the assumption that the non-hydrostatic equatorial bulge was a genuine characteristic of the planet. He suggested values for lower mantle viscosity which were in excess of 10^{27} Poise. Both of these arguments for high lower mantle viscosity were completely undermined by Goldreich and Toomre (1969) who pointed out that the non-hydrostatic equatorial bulge did not in fact exist! It had been inferred from a spherical harmonic expansion which was improperly biased towards the n = 2 harmonic.

At the same time new rebound data were becoming available from the Laurentide uplift which offered the possibility that the lower mantle viscosity could be deduced directly. As mentioned previously, the ice sheet over Canada covered an area so large that the dominant wavenumber for the relaxation is reduced by a factor of three from that for Fennoscandia. We have already seen that the relaxation time for

the central Laurentide uplift (about 1700 years) is roughly what one would expect if the viscosity in the middle to lower mantle were the same (under Canada) as that which exists in the upper mantle beneath Fennoscandia.

The implication of the above observation is that beneath the high viscosity lithosphere the variation of mantle viscosity with depth is not extreme. Before this interpretation can be regarded as reasonable, however, there are many possible objections which must be directly assessed. This requires a much more complete model of the isostatic adjustment process than the uniform one in terms of which most of the above discussion has been formulated. However, in addition to allowing for radial variations of physical properties, the model should be one which predicts the observed sea level variations themselves. We do not observe free relaxation times for separate wavelengths of the response directly; rather we observe the relative sea level histories at discrete points on the surface, variations which are forced by the simultaneous melting of specific large scale ice sheets. A complete model of the process must include the ice sheets and oceans explicitly and be capable of describing the gravitational interaction between them and the viscoelastic Earth on which they reside. The discussion in the following sections is intended to motivate a physical understanding of these three ingredients of the model and of the way in which they interact.

2. Viscoelastic Earth Models

The rheological behaviour of any material is described from a continuum mechanical point of view through a constitutive relation between applied stress and realized strain. From past and ongoing research in body wave and free oscillation seismology it has been amply demonstrated that for most purposes such phenomena can be described in terms of a Hookean elastic solid Earth. Even for deformations with such short characteristic time scales, however, this model is inadequate for some purposes. Since it includes no mechanism for the dissipation of contained elastic energy it can neither explain the finite "Q" of the elastic gravitational free oscillations nor the observed spatial attenuation of propagating surface waves. For much longer deformation time scales, on the order of the Earth's age, the planet deforms such that a stress of large spatial scale is completely relaxed and therefore appears to behave as a fluid. This may be argued on the basis of the observation that the Earth's shape is to a good first approximation in gravitational equilibrium with its current rotation rate (i.e. the stress associated with the centrifugal body force is balanced by the polar flattening and this was presumably accomplished by the fluid-like flow of matter in the interior of a planet whose shape was initially irregular). Because the time scales associated with isostatic adjustment are in some sense "intermediate" between the extremes of seismology and long time scale geological processes we require a rheological model of the interior which is "correct" in both limits.

The constitutive relation which has been employed in descriptions of the adjustment process satisfies this criterion and is the simplest model which is capable of doing so. It is linearly viscoelastic and in the large class of such rheological models the one we use is normally referred to as the Maxwell solid (Peltier, 1974). When it is subject to an applied shear stress such a solid initially behaves as if it were Hookean elastic. If the applied stress is maintained the material subsequently begins to behave as if it were a Newtonian viscous fluid. The Maxwell solid is characterized by 4 physical properties ρ, λ, μ, ν which are respectively the density, the elastic compressibility, the elastic rigidity, and the viscosity. In this model the time scale over which the transition from Hookean elastic to Newtonian viscous behaviour is effected is the Maxwell time $T_M = \nu/\mu$. The parameters ρ, λ, μ are common to the conventional Hookean elastic model of seismology and are known functions of radius determined by inversion of the free oscillations data (Gilbert and Dziewonski, 1975). The remaining parameter $\nu(r)$ is to be determined by fitting the model to the isostatic rebound data. Such models of material behaviour have been at least partially employed in several analyses of glacial isostasy (e.g. McConnell, 1968; Cathles, 1975; Peltier, 1974; Peltier and Andrews, 1976).

The strongest assumption in such models is that the stress-strain relation is linear and this is a subject of current contention. Recent experimental evidence (Ashby, 1972; Stocker and Ashby, 1973; Post and Griggs, 1973; Kohlstedt and Goetze, 1974) on the creep of mantle material (mostly Olivine single crystals) suggests that the creep rate limiting process is not linear. However, the laboratory data are necessarily taken at creep rates which are enormously in excess of those which are associated either with rebound or with convection $[0(10^{-6} \text{ s}^{-1})$ rather than $10^{-16} \text{ s}^{-1}]$. It is therefore unclear whether a linear mechanism might not obtain in the regime of lower creep rates. It has been argued furthermore, by Post and Griggs (1973), that the Fennoscandian data themselves suggest a non-linear stress-strain relation. This is based upon an assumption as to the magnitude of the negative gravity anomaly remaining over Fennoscandia. Their conclusion is very sensitive to this assumed magnitude, the error of which is liable to be large.

The point of view which we adopt here is that the ability of the simple linear viscoelastic models to fit the data should be tested carefully before such models are rejected. However, it must be kept clearly in mind that even if such models do "fit" this does not necessarily imply that the mantle is viscoelastically Newtonian. To show this would require an explicit demonstration that non-Newtonian models are incapable of providing similar accord with the observations. Brennan (1974) and Crough (1977) have completed some initial work on non-Newtonian models but much remains to be done in this area.

For the linear viscoelastic models there are two important ways in which we may summarize the manner in which they respond to an applied load on the surface. The first is in terms of a relaxation diagram (Peltier, 1976) which essentially gives the free decay time as a function of the spatial scale of a harmonic surface deformation. This is the same information as was employed for interpretation purposes in the Introduction. The second summary representation of the response is provided by the so-called Green function for the surface mass load boundary value problem. Physically, this is just the answer to the following question. How does the planet deform (we can describe this deformation in several ways; e.g. changes in radius, change in surface gravitational potential, change in surface gravitational acceleration, etc.) when it is "loaded" by the addition onto its surface of a unit point mass?

In Figure 4 we show four versions of the relaxation diagram for spherical Earth Models of increasing complexity. The model parameters are described in the Figure caption. Figure 4a is the relaxation diagram for an Earth model in which all of ρ, λ, μ and ν are constant and illustrates the relationship between relaxation time $\tau = s^{-1}$ and spherical wavenumber n = L discussed in the Introduction (all times are nondimensionalized with a characteristic scale of 10^3 years). For sufficiently large n, τ increases linearly with n in accord with the prediction of the equivalent halfspace model. In addition, there is one and only one relaxation time associated with each value of n. In Figure 4b we show the effect upon the relaxation diagram of including a lithosphere at the surface of the planet and for illustration the viscosity in this region has been taken to be infinite and the thickness equal to 120 km. Two effects are apparent. Firstly, the presence of the lithosphere reduces the relaxation times for the short deformation wavelengths (large n). Secondly, it introduces a completely new relaxation time for each value of n so that for each wavelength there are now two accessible modes. The two modes of relaxation have been labelled M0 and L0 on the Figure, the first corresponding to Mantle and the second to Lithosphere. The two modal lines coalesce at large n, a mathematical manifestation of the physically intuitive result that for sufficiently short wavelength all viscous gravita-

tional relaxation is supressed. Such short
deformation wavelengths are supported elasti-
cally. Note from this Figure that for n =
0(10) (Fennoscandia) and for n = 0(100) (Lake
Bonneville) the relaxation times for the dominant Mantle mode are both on the order of 5000
years. This is the basis of the remarks in the
Introduction in connection with Crittenden's
interpretation of the Lake Bonneville data.
In Figure 4c we show a relaxation diagram for

an Earth model with an inviscid high density
core, a constant viscosity mantle, and no
lithosphere. Again, the main effect of the
presence of the core is to introduce a second
relaxation time into the relaxation spectrum
for each value of n. Both the fundamental
mode of the mantle (MO) and the core mode (CO)
have relaxation time increasing as the wavelength decreases.

In Figure 4d we show the relaxation diagram

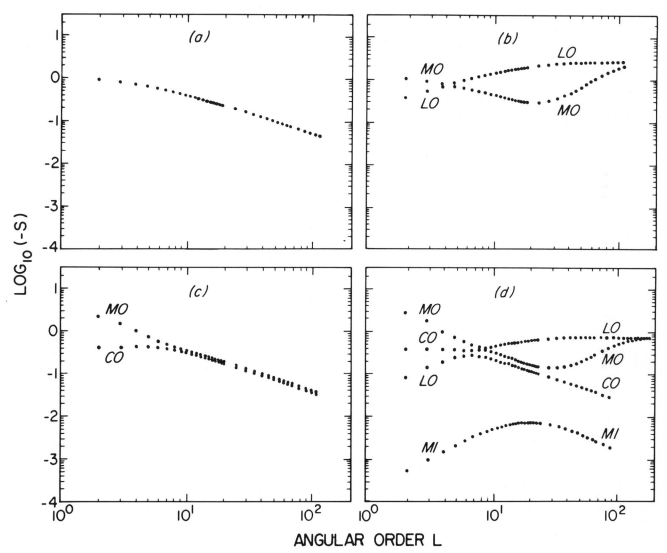

Figure 4: Relaxation diagrams for four earth models including various realistic effects. The
variable s in the logarithmic scale for the ordinate is the inverse of the relaxation time τ
(i.e. $s = 1/\tau$) and the time is scaled in units of 10^3 years. Thus the ordinate value 0 corresponds
to a relaxation time of 10^3 years, the value −1 to $\tau = 10^4$ years, etc. (a) is for an earth model
with ρ, λ, μ constant and viscosity $\nu = 10^{22}$ Poise. (b) is for the same earth model with a lithosphere of thickness 112.5 km in which $\nu = \infty$. (c) is the same as (a) but includes a high density
liquid core and does not include a lithosphere. The elastic structure of the core and mantle are
the same as the corresponding mean properties for the real earth. (d) is for a complete Earth
model whose elastic structure is taken from Gilbert and Dziewonski (1975). The model has a lithosphere which is 120 km thick and an inviscid core. The mantle viscosity is 10^{22} Poise.

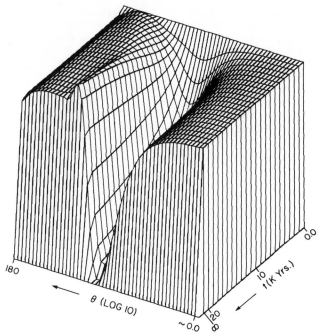

180

θ (LOG 10)

120

~0.0

0.0

10

t (K Yrs.)

0.0

Figure 5: Viscous part of the Green function
for radial displacement for the realistic Earth
model whose relaxation diagram was shown in
Figure 4(d). The Green function has been
scaled by multiplication with the factor $(a\theta)$
where a is the Earth's radius and θ is the angular separation from the point load on the
surface. This normalization removes the goe-
metric singularity at $\theta = 0$ for plotting
purposes.

for a complete Earth model. This has a λ, μ, ρ
structure which is identical to model 1066A of
Gilbert and Dziewonski (1975). In addition it
has a lithosphere which is 120 km thick and an
inviscid core. Throughout the mantle the viscosity is 10^{22} Poise. Inspection of this Figure
shows that the relaxation diagram contains each
of the three main modal branches discussed above
but there is an additional mode of relaxation
which has long relaxation time and which is labelled M1. This mode is supported by internal
density jumps in the mantle associated with the
Olivine-Spinel and Spinel-post Spinel phase
changes.

In the literature on isostatic adjustment it
is not generally appreciated that the relaxation
spectrum for realistic Earth models has the
richness described above. Each wavelength in
the relaxation has several relaxation times accessible to it. For a given wavelength, the
importance of each mode is determined by the
"efficiency" with which it is "excited" by an
applied surface load of the same spatial wavelength. The actual scale of a realistic surface load (ice or water) determines which wavelengths (values of n) will dominate the res-

ponse forced by its application, but has nothing
to do with the relative importance of the various relaxation times which are accessible to
each n. This depends upon the mechanism by
which the relaxation is forced and in glacial
isostasy the mechanism of forcing is through
the gravitational interaction between the surface mass load and the planet itself.

Given the relaxation diagram, we may proceed to determine the way in which the corresponding Earth model deforms in response to a
specific applied load. It is convenient to
consider first the response produced by the application to the surface of a point load of unit
mass. This response is called the Green Function
for the gravitational interaction problem and
its importance relies upon the fact that the
Earth model which we employ has a linear rheology. Because the model is linear we may invoke
the principle of superposition and the response
of the planet to an arbitrary space and time dependent surface load may be calculated simply
by "adding up" the response due to the individual
(small) component parts into which it may be
decomposed.

For a Maxwell model it has previously been
shown (Peltier, 1974) that the Green Function
may be separated into the sum of two distinct
parts, an elastic part which contains the in-
stantaneous elastic deformation, and a viscous
part which contains the viscous relaxation. In
Figure 5 we show the viscous part of the Green
Function for radial displacement for a realistic
Earth model which has a lithosphere of thickness
120 km in which $\nu = \infty$, a high density core in
which $\nu = 0$, and a mantle in which $\nu = 10^{22}$
Poise (c.g.s.) everywhere. The Green Function
depends upon the angular separation θ between
the point load and the position on the surface
at which the response is determined and upon
the time t. For plotting purposes the actual
function has been normalized by multiplication
with (a x θ) where "a" is the radius of the
Earth. At time t = 0 when the point mass is
applied the viscous part of the response is
zero. As time progresses the surface begins to
sink under the load, at first quickly and then
more slowly. In the limit of infinite time a
new state of isostatic equilibrium is reached
in which the new shape of the planet is in
gravitational equilibrium with the point load
on its surface. By inspection of Figure 5 you
can see that at sufficiently great distances
from the load the local radius has actually
increased. It is the collapse of this "peripheral bulge" if the load is subsequently
removed, which explains the characteristic
donut-shaped response within and surrounding
deglaciated regions which was shown previously
in Figure 3. The local radius increases in regions which were under the ice and decreases in
the periphery due to the collapsing bulge
(Peltier, 1974).

It therefore appears that the simple linear

viscoelastic Earth model does contain all of the necessary characteristics which it must have if it is to fit the gross characteristics of the relative sea level data set. We may describe the behaviour of such Earth models in terms of several different signatures of their response to a point load. For example, we may calculate the gravity anomaly on the deformed surface or the local surface tilt (Peltier, 1974). The model may therefore be employed to predict these additional observables. We may also calculate a Green Function for the perturbation of the gravitational potential on the deformed surface. In Section 4 we shall show that the latter may be employed to calculate relative sea level directly.

3. Models of Surface Deglaciation and Initial Estimates of Rebound

In order to use the above described Green Functions for isostatic adjustment to predict the observed variations of relative sea level, present day gravity anomalies, etc., we require knowledge of the actual loading history of the surface. These surface loads comprise two distinct parts associated respectively with the ice sheets and with the meltwater they produce. Clearly, these two components are related by the requirement that they conserve mass. The negative load applied to the glaciated regions when the ice sheets melt appears subsequently as meltwater in the ocean basins which are therefore subject to a positive load. As we will show in Section 4 the structure of the global problem is such that given the deglaciation history we may determine the history of the ocean loading simultaneously with the calculation of relative sea level curves. Here we describe briefly the problem of reconstructing the northern hemisphere deglaciation chronology.

There are three distinctly different types of information which are required to reconstruct the disintegration histories of the major ice sheets. Together, they lead to a time sequence of maps of ice sheet thickness during deglaciation. We begin by estimating from the observations of relative sea level themselves the total mass which must have been contained in all ice sheets as a function of time. This estimate is obtained by inspection of the relative sea level record from regions located at great distance from the ice sheets. The data from such reions are only weakly affected by variations in the local radius associated with either the elastic or with the viscous response to load removal. If we assume that the meltwater added to the oceans is distributed uniformly over the entire area of the ocean basins then knowing this area and having observed the time dependent submergence at distant sites we may estimate the increase in ocean volume due to the melting as a function of time. It is then

simple to calculate the mass which the ice sheets must have contained to produce the observed worldwide sea level rise. In Figure 6 we may compare the mass loss history of the disintegration model described in Peltier and Andrews (1976) and Andrews and Peltier (1976) to the observations of Shepard (1963). The model fits this integral constraint quite accurately.

This information by itself is insufficient to determine ice sheet thickness as a function of place and time. We may obtain a first approximation to the actual spatial distribution by invoking two additional kinds of data. From the location and age of glacial terminal morains it is possible to reconstruct a sequence of maps showing the areas within which the ice masses were found as a function of time. The total mass contained within each ice sheet is then determined by partitioning the total mass among the ice sheets in proportions determined by their relative areas. The topography of the ice sheets may be estimated by invoking quasi-steady state ice mechanical arguments (e.g. Patterson, 1972) and additional field evidence. In this way we may obtain a first estimate of the ice unloading history for each of the two major ice regions, Fennoscandia and Laurentide. In Figure 7 (a,b,c) we show maps of the thickness of Laurentide ice for three times during the unloading history tabulated by Peltier and Andrews (1976). The same time sequence for Fennoscandia is shown in plates d,e, and f. It must be kept clearly in mind that these reconstructions of the melting histories are

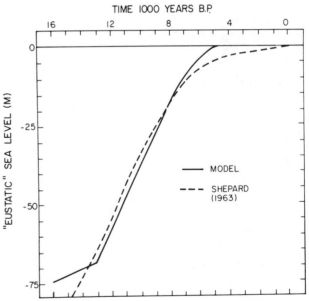

Figure 6: Comparison of the model eustatic curve based upon the Peltier and Andrews (1976) deglaciation history with the observed curve of Shepard (1963).

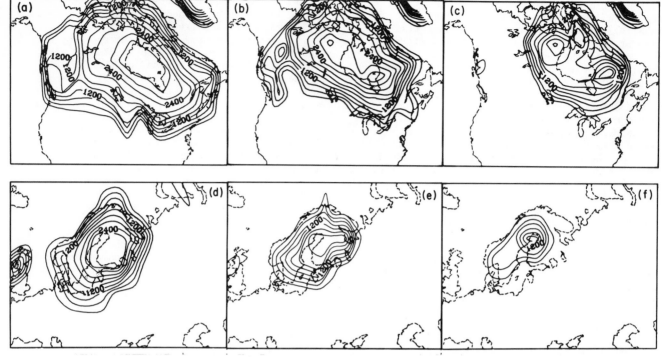

Figure 7: Plates (a), (b), and (c) show three time slices through the ICE-1 melting history at 18 ka BP, 12 ka BP, and 8 ka BP respectively for the North American ice complex. Plates (d), (e), and (f) show the same sequence for Fennoscandia.

first approximations which are unlikely to be correct in detail. They will have to be refined in order to improve the fit of the theory with the observed variations of relative sea level.

This brings squarely into focus an important characteristic of theoretical models of glacial isostasy. Such models require two distinctly different inputs before they can be employed for prediction: (1) a rheological model of the interior, the anelasticity of which is characterized by an effective viscosity $\nu(r)$, and (2) a model of ice sheet disintegration. We attempt to determine both by fitting the model to the observations. The problem of inference is therefore a highly non-linear one. It is our ability to perform an initial linearization of this problem by using *a priori* information to reconstruct the disintegration histories of the major ice sheets which allows us to partially separate the problem of inferring viscosity from the problem of determining the load history. We attempt to determine these two unknown "functionals of the model" by an iterative process described by the formal inverse theory in Peltier (1976).

We may check the first order validity of our ice sheet reconstructions by calculating approximate sea level histories for a few points on the surface which are in the "near field" of the ice sheets so that the effect of ocean loading by

meltwater may be neglected to first order (Peltier, 1980a). In addition we will make a second strong assumption which has been traditional in such work but which may lead to significant error as will be discussed in the next section. We shall assume that the relative sea level data in the post-deglaciation phase accurately reflect, and only reflect, changes in the local radius of the planet. The geoid (surface of the ocean) is therefore assumed to be located at a fixed radius from the centre of mass of the planet at each point on the surface after deglaciation. Based upon these assumptions we may predict relative sea level in the post deglaciation period by decomposing each ice sheet melting history into a sequence of point load removals. Then, using the Green Function for radial displacement to determine the response to each point load we simply add up the response due to all point loads in space, appropriately delayed in time, to determine the relative sea level history at a specific surface location.

The results from such simplified calculations have been discussed in detail by Peltier and Andrews (1976) who suggested on the basis of them that Earth models with high lower mantle viscosity did not fit the observed relative sea level variations along the east coast of North America. This result was also obtained by Cathles (1975) who employed a model based upon the same physical assumptions but a different theoretical formalism. It has had some impact upon the debate concerning the **style** of mantle convection which is responsible for the motion of surface plates since a uniform mantle viscosity strongly favours the notion of whole mantle as opposed to upper mantle convection. Quantitative arguments in favour of the whole mantle model are found in Peltier (1972), Sharpe and Peltier (1978), and Peltier (1980b).

Recent calculations employing more complete models of the isostatic adjustment process (Peltier et al., 1978, Clark et al., 1978; Peltier and Wu, 1980) have not, however, served to completely confirm the preliminary strong conclusion which was based upon the simplified models. This new work has revealed a tradeoff of sorts between errors made in inferring the load history and errors made in inferring the viscosity profile. A detailed discussion of this tradeoff is provided in Peltier and Wu (1980) and the next section of this paper will summarize the results of this new work. The problem of imperfect knowledge of the unloading history is not the only inadequacy of the Peltier and Andrews (1976) analysis. In their work ocean loading was neglected entirely and the geoid was assumed stationary. The new models incorporating these physical interactions are discussed below as are the new insights concerning the inverse problem for mantle viscosity which have been obtained through its' application.

4. Gravitationally Self Consistent Sea Level Variations on a Deglaciating Viscoelastic Earth.

Both of the difficulties with the simplified rebound model enumerated above are connected with a problem mentioned earlier. When the ice sheets melt and their meltwater enters the ocean basins, we must be able to determine where in the oceans the water accumulates. Only then will it be possible to describe the differential ocean loading accurately and only then will we know how the geoidal surface is deformed. It has been conventional in the literature of glacial isostasy to answer this question in a particularly simple fashion (e.g. Cathles, 1975). One assumed that the added meltwater was spread uniformly over the entire surface of the oceans as a function of time so that the variation of bathymetry was the same everywhere. This assumption is clearly incorrect and in fact leads to errors of prediction which may be significant in some locations. The reason for this is physically clear. The equilibrium surface of the global ocean (the geoid) is of necessity a surface on which the gravitational potential is constant. If for any reason the potential on this surface is locally perturbed then the resulting unbalanced gravitational force will produce currents in the water which will redist-

ribute the water mass such that constancy of the surface potential is restored.

If we assume, and this is an important assumption, that during glacial maximum at approximately 20 ka BP, the ice sheets, oceans, and solid Earth were in a state of gravitational (isostatic) equilibrium, then we may envision the following scenario as the ice sheets melt. Initially the sea level is held anomalously high in the vicinity of the ice sheets by the direct gravitational attraction of the water by the ice. However, everywhere on the surface of this initial ocean the gravitational potential is a constant since it is in equilibrium. When melting commences the potential on the surface is perturbed non-uniformly and the added meltwater is distributed over the oceans such as to restore the potential to a new constant value everywhere. In response to the load added over the ocean basins the sea floor will be depressed and in response to the load removed where the ice sheets are melting the land will be elevated. The complex redistribution of mass in the interior of the planet which is effected by the net load variation will force further irregular variations of potential on the ocean's surface and thus further redistribution of water will be required to equalize the potential. This process of continual gravitational "feedback" between the ice sheets, the oceans, and the solid earth is the process which ultimately determines the relative sea level signature which will be observed everywhere where continent and ocean meet. In order to accurately predict the sea level variations the entire sequence of interactions must be described in detail.

Although these feedback processes are complicated they may be described completely using the Green Function formalism for the isostatic adjustment problem developed in Peltier (1974). The key ingredient in the theory is the Green Function for the perturbation of the ambient gravitational potential on the deformed surface. Given a model deglaciation history, and this functional of the viscoelastic Earth (e.g. Peltier and Andrews, 1976; Peltier et al., 1978), we may construct an explicit integral equation for the relative sea level variation everywhere on the Earth's surface. The derivation of this sea level equation is described in detail in Farrell and Clark (1976) and in Peltier et al. (1978). The solution of the sea level equation gives a direct prediction of the time dependent vertical separation between the surface of the ocean and the surface of the solid Earth and therefore may be directly compared to the relative sea level data. Both the ocean loading and the deformation of the geoid are explicitly accounted for. Peltier (1980a) has given a detailed discussion of the relation between this new theory and the conventional one in which geoidal variations are neglected and the ocean loading is taken as spatially uniform.

There it is shown that the previous theory is valid only in the near field of ice sheets which are sufficiently small and for times after melting is complete. For sufficiently large ice sheets the previous theory may be in error even in the near field. It will also make incorrect relative sea level predictions in the near field of all ice sheets during the deglaciation phase and will always be in error at sufficiently great distances from the regions of active deglaciation. In general, we must therefore conclude that geoidal variations may have an important direct effect on the sea level record. In retrospect, this should not appear surprising.

An example of a global solution of the gravitationally self consistent sea level equation is shown in Figure 8 where the relative sea level variation is contoured in metres for 4 times following deglaciation. The Earth model employed has radial variations of elastic parameters and density which fit the free oscilla-

tion data and has a viscosity profile in which $\nu = 10^{22}$ Poise (cgs) between the Earth's surface and the core mantle boundary and an inviscid core. It is the model which preliminary calculations suggested to provide a good fit to the relative sea level data. Inspection of Figure 8 shows that the relative sea level variations are highly non-uniform over the oceans, even in the far field of the ice sheets, as must be the case if the ocean surface is to remain an equipotential. Over the Hudson Bay and Sweden, which were the centres for the Laurentide and Fennoscandian ice sheets, strong decreases of relative sea level obtain which are dominated by the local uplift of the surface of the solid Earth. The "output" from the solution of the sea level equation is a sequence of maps such as those shown in Figure 8 which are spaced arbitrarily in time. In all of our calculations to date we have elected to sample the response at equally spaced times separated by 10^3 years. To compare the theoretical predictions with obser-

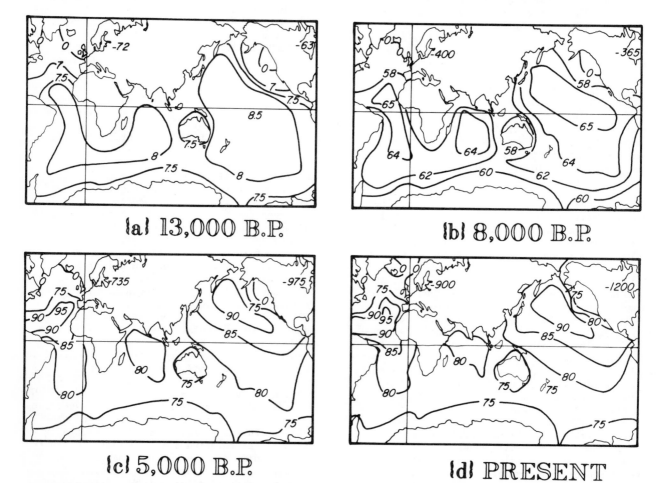

Figure 8: The global rise of sea level (in m.) at four times subsequent to the onset of melting. Note the large negative values corresponding to a fall of local sea level in the vicinity of the Laurentide and Fennoscandia ice sheets. The rise of sea level is not uniform in the far field showing explicitly that the concept of eustatic sea level is of limited utility.

vation we simply "stack" the results for a particular point on the surface at which data are available, properly phased in time, and overlay the prediction with the observation.

On the basis of a global analysis of the relative sea level signature predicted by the model we may divide the entire surface of the Earth into six regions, in each of which the history of relative sea level is distinctive of that region. These are shown in Figure 9 where they are labelled I-VI (Clark et al., 1978; Peltier et al., 1978). The characteristics of the sea level record in each zone and the physical explanation of these characteristics are as follows. In I the sea level record is monotonic and shows the continuous fall of sea level in regions which were under the ice. In such regions the sea level record is dominated by the uplift of the surface of the solid Earth which at first proceeds quickly and later slows, the time history being roughly exponential as illustrated previously by the uplift curve in Figure 2. In region II the sea level record is again monotonic but here the sense of the variation is reversed and continuous submergence is expected. This pattern, with a region of submergence surrounding the region of central emergence, is in accord with observation (see Figure 3). On the boundary between regions I and II there is a thin zone which is not shown on Fig-

ure 9 in which the sea level history is not monotonic - initial emergence is followed by submergence. Such effects are also seen clearly in the data for the east coast of North America discussed by Walcott (1972b). These non-monotonic histories are produced by the propagation of the peripheral bulge after removal of the glacial load as discussed in Peltier (1974). Region III is also narrow and in it the model predicts a very slight amount of emergence at present. In region IV no raised beaches should be observed and continuous submergence again dominates the sea level record. Region V has the interesting characteristic that raised beaches begin to form as soon as glacial melting is complete. In this region the local radius of the planet is increasing continuously during and subsequent to melting, a fact which can be seen by direct inspection of the Green Function of the system. While the water is being added to the oceans, however, the rate of increase is insufficient to "keep pace" with the increase in water depth so that no raised beaches can form. Once melting is complete the local bathymetry becomes approximately stationary and the land rises above the sea surface and new beaches are cut, the old now lying above the geoid. Region VI consists of all continental shorelines which are sufficiently distant from the deglaciation centres. In these areas raised beaches are

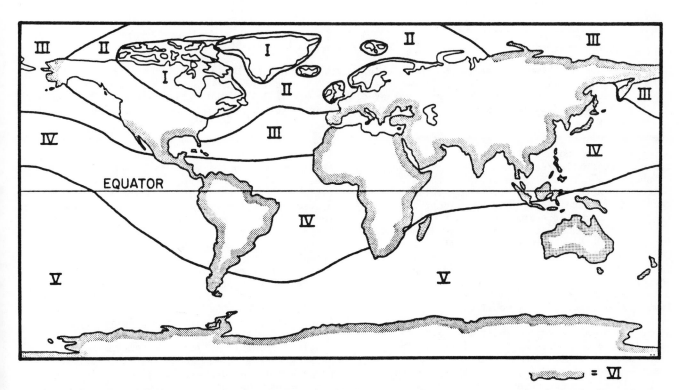

Figure 9: The global extent of six regions labelled I-IV in the Figure, in each of which the relative sea level curve has a certain characteristic signature. The form of each of these characteristic signatures is described in the text.

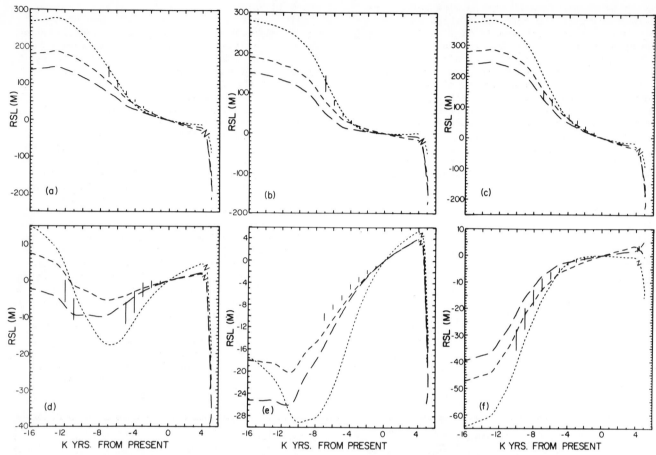

Figure 10: RSL histories at (a) Churchill, (b) Ottawa Islands, (c) Cape Henrietta Maria, (d) Boston, (e) Clinton, Connecticut, (f) Bermuda for the three different viscosity models discussed in the text. The observed data are represented by the vertical bars. The final dramatic excursion of each RSL prediction ends at the point which corresponds to the amount of emergence (submergence) remaining.

produced by the upward tilting of the land in response to the increased waterload immediately offshore. The effect is measurable and is completely in accord with observation. This effect was previously described by Walcott (1972a) in terms of a simple model in which the effects of ice and water loading were decoupled. The model employed here is fully self consistent gravitationally and so all physical processes are simultaneously included and properly accounted for. The model is valid globally.

A systematic comparison of the relative sea level predictions of this model with the observed sea level variations at a global set of sites has been documented fully in Clark et al. (1978) and in Peltier et al. (1978). The existence of the six zones predicted by the model was established and, particulary in the far field of the ice sheets, the fit of the model to the observations was rather good. As discussed in Peltier (1980a), the fit to the far field data was obtained at the sacri-

fice of the agreement between theory and observation at sites which were under the ice sheet. This was achieved in Clark et al. (1978) by using the disintegration history tabulated in Peltier and Andrews (1976) shifted forward 2000 years in time. This 2000 year time shift was controlled only by the time of appearance of raised beaches in zone VI. It is possible to improve the fit of the theory to near field sites, in the context of the time shifted version of the Peltier and Andrews load history, by reducing the thickness of the Laurentide ice sheet. The ability of this modified load history to reconcile the relative sea level data will of course be a function of the mantle viscosity profile. In Peltier and Andrews (1976) it was shown that the rsl data preferred a constant mantle viscosity of about 10^{22} Poise when the original disintegration model was employed.

In Figure 10 we illustrate comparisons of observed and predicted rsl histories at six

North American sites for three different viscosity profiles which differ only in the magnitude of the viscosity beneath 670 km depth. These calculations used the modified load history. The first three sites, which are all around Hudson Bay near the centre of rebound, are (a) Churchill, (b) Ottawa Islands, and (c) Cape Henrietta Maria. At all three sites the data reject the model with the highest lower mantle viscosity of 5×10^{23} Poise (long dashed line). There is no unanimous preference for either of the remaining two models (the dotted curves are for the uniform 10^{22} Poise model while the short dashed curves are for the model with lower mantle viscosity of 10^{23} Poise) although either fits these data reasonably well. Inspection of similar comparisons at the remaining sites (d) Boston, (e) Clinton Connecticut, and (f) Bermuda, which were all outside the ice margin, demonstrates that some increase of lower mantle viscosity is required to fit the submergence data in this region (e and f in particular). Again the model with highest lower

mantle viscosity is everywhere rejected.

The few comparisons of observed and predicted rsl histories described here suffice to demonstrate the nature of a tradeoff between errors in the load history and errors in the viscosity profile which is fundamental to the viscosity inverse problem. A more comprehensive discussion will be found in Peltier and Wu (1980). Using the ICE-1 deglaciation history tabulated in Peltier and Andrews (1976), the rsl data prefer a viscosity profile in which the lower mantle viscosity is not much more than a factor of two or so higher than the upper mantle value near 10^{22} Poise. With the modified deglaciation history (called ICE-3) in which the Laurentide ice sheet is thinner than in ICE-1 and melts 2000 years later, the rsl data require about an order of magnitude increase in lower mantle viscosity. Although we are continuing to investigate the ambiguity in the inference of viscosity allowed by the range of plausible load histories, it should be kept clearly in mind that the delayed melting characteristic of ICE-3 violates

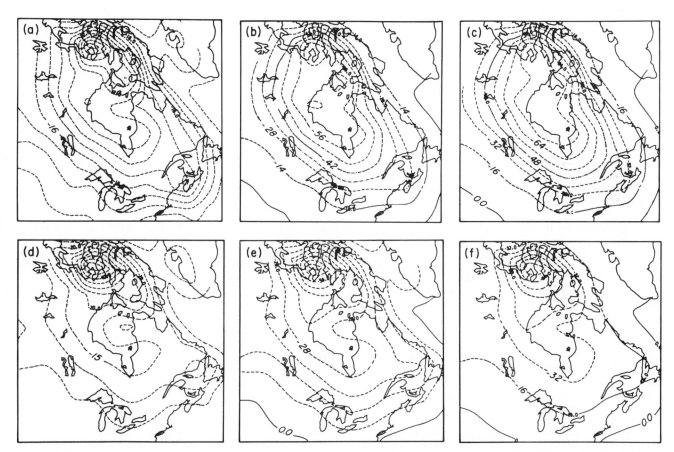

Figure 11: Free air gravity predictions for the Laurentide region. Plates (a), (b), and (c) show results from calculations based upon ICE-1 for the uniform viscosity model, that with a lower mantle viscosity of 10^{23} Poise, and that with a lower mantle viscosity of 5×10^{23} Poise. Plates (d), (e), and (f) show the same sequence of calculations but based upon the ICE-3 disintegration history described in the text.

the observed disintegration isochrones. To the extent that ICE-1 is the preferred melting history the viscosity of the deep mantle must be reasonably low. In the next section we shall consider the extent to which this conclusion is compatible with the observed free air gravity anomalies over the formerly ice covered regions.

5. The Free Air Gravity Anomaly in the Central Depression.

The importance of the free air gravity anomaly to the problem of inferring the viscosity of the mantle is to be found in the fact that these data provide information which is complementary to that obtained from the rsl histories. Because the free air signal is a measure of the extent of existing isostatic disequilibrium,it is sensitive only to the amount of uplift remaining. The rsl histories, on the other hand, essentially record the amount of uplift or submergence which has already been realized. Gravity data "looks into the future" whereas sea level histories "look into the past" so that the former are most influenced by the longest relaxation times in the deformation spectrum whereas the latter are most influenced by the shortest relaxation time.

The free air gravity anomaly is a signature of the isostatic adjustment process which may be calculated using the same Green function formalism employed to simulate rsl histories. The solution of the sea level equation provides a history of the local bathymetry variations in the global ocean and when this is coupled with the input model of ice sheet disintegration we then have a complete surface mass loading history since deglaciation. This may then be convolved with the Green function for the free air anomaly (Peltier, 1974) to produce theoretical predictions of the time dependent gravity anomaly anywhere on the Earth's surface. We can in fact compute an array of such predictions on a grid and present the prediction as a contoured map of the free air signal for a particular location. Clearly we shall be most interested in the locations which were once ice covered since in these regions the anomaly, being largest, is least likely to be obscured in "noise" derivative of other geological processes.

In Figure 11 we compare a sequence of present day free air gravity maps calculated for the Laurentide region. The first three plates show calculations for the ICE-1 load history for each of the previously mentioned viscosity models. Plate (a) is for the model with a uniform mantle viscosity of 10^{22} Poise, (b) for the model with a viscosity of 10^{23} Poise beneath a depth of 670 km, and (c) is for the model with a lower mantle viscosity of 5×10^{23} Poise. Plates (d), (e), and (f) are for the same sequence of viscosity models but with the ICE-3 load history. The observed free air signal for this region is

shown in Walcott (1970) and consists of a ellipsoidal anomaly centred upon Hudson Bay whose major axis is aligned in a North Westerly direction. The minimum anomaly is approximately -35 mgals and obtains approximately over the Hudson Bay itself. Since this regions is a sedimentary basin, and since such regions normally have positive free air anomalies associated with them, the actual anomaly due to glacial isostatic disequilibrium could be as large as -45 mgals. That the edge of the observed negative anomaly, as determined by the zero anomaly contour, is virtually coincident with what was the edge of the ice sheet, is rather strong evidence of a glacial origin.

Inspection of the computed anomalies shown in plates (a), (b), and (c) strongly reinforces the validity of the viscosity model inferred from the rsl data (Peltier and Andrews, 1976). The present day free air anomaly calculated for the uniform viscosity model is approximately -35 mgals, which is equal to the observed anomaly. The anomalies for the models with high lower mantle viscosity are both in excess of -60 mgals, so that these models are both absolutely rejected by the gravity data. In terms of the ICE-1 load history of Peltier and Andrews (1976) even a lower mantle viscosity of 10^{23} Poise, only one order of magnitude higher than the viscosity of the upper mantle, is strongly rejected by the gravity data. This is of course subject to the assumption that the ice sheets were in isostatic equilibrium initially (Peltier, 1981; Peltier and Wu, 1980). The calculations shown in plates (d), (e), and (f) for the ICE-3 load history also reinforce the conclusion based upon the rsl predictions for this model. The uniform viscosity model predicts a negative gravity anomaly over Hudson Bay of only (-15) - (-20) mgals or half that which is observed. From plate (e), increasing the lower mantle viscosity to 10^{23} Poise gives an anomaly of about -32 mgals which is marginally acceptable while a further increase to 5×10^{23} Poise (plate f) yields approximately -36 mgals which is a slight further improvement. Therefore, if ICE-3 is nearer the correct load history then a significant increase of viscosity across the 670 km discontinuity is required to fit both the gravity and the rsl data. It is worth reiterating here the fact pointed out at the end of the last section, however, that ICE-3 does not fit the known disintegration isochrones for the Laurentide sheet. Considering therefore the totality of the observation data, the uniform viscosity model is preferred.

6. Conclusions

In the course of the previous discussion we have tried to trace the recent developments of models designed to simulate the phenomena associated with glacial isostatic adjustments. It is now well established, in my opinion, that linear models of this process are quite capable of fitting the observations. Such models require an upper mantle viscosity of approximately 10^{22} Poise and a lower mantle viscosity which is somewhat higher but definitely less than 10^{23} Poise. The uniqueness of this inference is currently under investigation using the formal inverse theory developed in Peltier (1976).

The main result which is to be derived from the calculations which have been completed to date is worth stressing as a final conclusion. The magnitude of the number obtained from studies of postglacial rebound for the mean viscosity of the mantle is compatible with the convection hypothesis of the origin of continental drift and sea floor spreading (Peltier, 1980). The observed present day value of the mean mantle viscosity may in fact be employed as a constraint upon thermal history models of the Earth since these are dominated by the mantle convective heat transfer (Sharpe and Peltier, 1978, 1979).

References

Andrews, J.T., and W.R. Peltier, Collapse of the Hudson Bay ice center and glacio-isostatic rebound, Geology, V. 4(2), 73, 1976.

Ashby, M.F., A first report on deformation mechanism maps, Acta Met., 20, 887, 1972.

Brennan, C., Isostatic recovery and the strain rate dependent viscosity of the earth's mantle, J. Geophys. Res., 79, 3393, 1974.

Cathles, L.M., The viscosity of the earth's mantle, Princeton University Press, New Jersey, 1975.

Clark, J.A., Global sea level changes since the last glacial maximum, unpub. Ph.D. thesis, Dept. of Geological Sciences, Univ. of Colorado, Boulder, Colo., 1977.

Clark, J.A., W.E. Farrell, and W.R. Peltier, Global changes in postglacial sea level: a numerical calculation, Quaternary Research, 9, 1978.

Crittenden, M.D., Effective viscosity of the earth derived from isostatic loading of Pleistocene Lake Bonneville, J. Geophys. Res., 68, 5517, 1963.

Crough, S.T., Isostatic rebound and power law flow in the asthenosphere, Geophys. J.R. astr. Soc., 50, 723, 1977.

Farrell, W.E., Earth tides, ocean tides, and tidal loading, Phil. Trans. Roy. Soc. London, Ser. A, 274, 45, 1973.

Farrell, W.E., and J.A. Clark, On postglacial sea level, Geophys. J.R. astr. Soc., 46, 647, 1976.

Gilbert, F., and A. Dziewonski, An application of normal mode theory to the retreival of structural parameters and source mechanism from seismic spectra, Phil. Trans. R. Soc. London A., 278, 187, 1975.

Goldreich, P., and A. Toomre, Some remarks on polar wandering, J. Geophys. Res., 74, 2555, 1969.

Haskell, N.A., The motion of a viscous fluid under a surface load, 1, Physics, 6(8), 265, 1935.

Haskell, N.A., The motion of a viscous fluid under a surface load, 2, Physics, 7(2), 56, 1936.

Haskell, N.A., The viscosity of the asthenosphere, Am, J. Sci., 33(193), 22, 1937.

Hays, J.D., J. Imbrie, and N.J. Shackleton, Variations in the earth's orbit, Pacemaker of the ice ages, Science, 194, 1121, 1976.

Heiskanen, W.A., and F.A. Vening Meinesz, The Earth and its gravity field, McGraw-Hill, New York, 1958.

Hillaire-Marcel, C., and R.W. Fairbridge, Isostasy and Eustasy in Hudson Bay, Geology, Jan., 1978.

Kohlstedt, D., and C. Goetze, Low-stress high-temperature creep in Olivine single crystals, J. Geophys. Res., 79, 2045, 1974.

McConnell, R.K., Viscosity of the mantle from relaxation time spectra of isostatic adjustment, J. Geophys. Res., 73(22), 7089, 1968.

McKenzie, D.P., The viscosity of the lower mantle, J. Geophys. Res., 71, 3995, 1966.

McKenzie, D.P., The viscosity of the mantle, Geophys. J.R. astr. Soc., 14, 297, 1967.

McKenzie, D.P., The geophysical importance of high temperature creep, in The history of the Earth's crust, ed. R.A. Phinney, Princeton University Press, Princeton, New Jersey, 1968.

Munk, W.H., and J.F. MacDonald, The rotation of the Earth, Cambridge University Press, New York, 1960.

Niskanen, E., On the viscosity of the earth's interior and crust, Ann. Accad. Sci. Fenn., Ser. A, 3(15), 22, 1948.

Parsons, B.E., Changes in the earth's shape, unpub. Ph.D. thesis, Cambridge University, Cambridge, England, 1972.

Paterson, W.S.B., Laurentide ice sheet, Estimated volumes during late Wisconsin, Rev. Geophys. Space Phys., 10, 885, 1972.

Peltier, W.R., Penetrative convection in the planetary mantle, Geophys. Fluid Dyn., 3, 365, 1972.

Peltier, W.R., The impulse response of a Maxwell Earth, Rev. Geophys. Space Phys., 12, 649, 1974.

Peltier, W.R., Glacial isostatic adjustment II: The inverse problem, Geophys. J.R. astr. Soc., 46, 669, 1976.

Peltier, W.R., Ice sheets, oceans, and the earth's shape, in Earth Rheology, Isostasy, and Eustasy, ed. N.A. Morner, John Wiley and Sons, Chichester, 1980a.

Peltier, W.R., Mantle convection and viscosity, in Physics of the Earth's Interior, ed. A. Dziewonski and E. Boschi, Elsevier North Holland, New York, 1980b.

Peltier, W.R, Ice age geodynamics, Annual Review of Earth and Planetary Sicences, in Press, 1981.

Peltier, W.R., and J.T. Andrews, Glacial-isostatic adjustment-I. The forward problem, Geophys. J.R. astr. Soc., 46, 605, 1976.

Peltier, W.R,. W.E. Farrell, and J.T. Clark, Glacial isostacy and relative sea level, a global finite element model, Tectonophysics, 50, 81, 1978.

Peltier, W.R. and P. Wu, Glacial isostasy, relative sea level, and the free air gravity anomaly as a constraint on deep mantle viscosity, Geophys. J.R. astr. Soc., submitted, 1980.

Post, R., and D. Griggs, The earth's mantle: evidence of non-Newtonian flow, Science, 181, 1242, 1973.

Sharpe, H.N., and W.R. Peltier, Parameterized mantle convection and the Earth's thermal history, Geophys. Res. Lett., 5, 737, 1978.

Sharpe, H.N., and W.R. Peltier, A thermal history model for the earth with parameterized convection, Geophys. J.R. astr. Soc., 59, 171, 1979.

Shepard, F.P., Thirty-five thousand years of sea level, Essays in Marine Geology, University of Southern California Press, 1, 1963.

Stocker, R.L., and M.F. Ashby, On the rheology of the upper mantle, Rev. Geophys. Space Phys., 11, 391, 1973.

Vening-Meinesz, F.A., The determination of the earth's plasticity from postglacial uplift of Fennoscandia: Isostatic adjustment, Proc. Kon. Ned. Akad. Wetensch, 40, 654, 1937.

Walcott, R.I., Isostatic response to loading of the crust in Canada, Can. J. Earth Sciences, 7(2), 716, 1970.

Walcott, R.I., Past sea level, eustasy, and deformation of the Earth, Quaternary Res., 2, 1, 1972a.

Walcott, R.I., Late Quaternary vertical movements in eastern North America: Quantitative evidence of glacio-isostatic rebound, Rev. Geophys. Space Phys., 10, 849, 1972b.

INSTRUMENTAL OBSERVATIONS OF MOVEMENTS AND STRESS
IN PLATE INTERIORS: INTRODUCTION

William M. Kaula

Department of Geophysics and Space Physics, University of California at
Los Angeles, Los Angeles, California 90024

The inferences discussed in Parts I, II and III depend on geological observations supplemented by radiometric measurements. Consequently, the data record for movements in plate interiors is incomplete in several ways: in variables measured, in geographic extent and in coverage of the temporal spectrum. Thus, for example, the measurement of the post-glacial rebound which occurred before modern times is largely confined to Carbon-14 datable shorelines, while stratigraphy unavoidably averages out most fluctuations of less than a 1000-year timescale.

Extrapolation from the geologic record suggests that the rates of motion in the solid earth should be observable by accurate modern techniques. Hence it is desirable that such observations be undertaken, not only to confirm the inferences from geologic data, but also to fill gaps in the data coverage and to determine the nature of shorter term motions. Earthquakes are the most obvious type of short-term motions; however, both observation and reasoning indicate that a wide variety of strain accumulation and creep phenomena in the solid earth should be measurable by modern techniques.

Measurements pertaining to solid earth tectonics include:

- stress: hydrofracture, overcoring
- change-of-stress: seismic motions
- rate-of-strain: strainmeter, tiltmeter, leveling, triangulation and trilateration
- gravitational acceleration: gravity meter.

The most direct are the overcoring and hydrofracture techniques, which are described in this volume by Gay. Hydrofracture has the advantage of measuring stress directly and of reaching to greater depths, thus avoiding relatively superficial noise. Measurement of the change-of-stress by seismic motion analysis has low precision and is somewhat model dependent. Measurements of rates-of-strain by strainmeters and tiltmeters have the advantage of continuous monitoring, but the severe disadvantage of sensitivity to local near surface effects. The geodetic techniques -- leveling, triangulation and trilateration -- average out local effects, and thus should yield results of broader interest. However, they are expensive and hence have been done only at infrequent intervals of some years (other than for special geodetic networks in plate margin earthquake zones). Gravity meters suffer the double disadvantage of the temporal changes (other than tides) being at the threshold of sensitivity, and the effect of local hydrological changes often being substantial, unless the measurement sites are located on crystalline rock outcrops.

The measurement difficulties described above result in the bulk of the applicable results today for plate interiors being either the relative elevation changes inferred from leveling, as discussed in the following article by Brown and Reilinger, or the stress inferred from hydrofracture and overcoring, as discussed by Gay. However, the prospects for obtaining new results from improved geodetic measurement techniques are good, as discussed by Bender.

The theme of the article on leveling results is that changes of elevation are measured which in many regions are at rates appreciably greater than indicated by geological evidence. Some of these are in areas of recent tectonic activity, and occur in the expected directions: e.g., uplifts in young mountains, subsidences in basins. However, the occurrence of rapid rates in some areas without evidence of recent tectonic activity, as well as discrepancies of leveling lines from each other and from tidal meters, make it desirable that these results be examined carefully for systematic errors before being accepted as evidence of tectonic activity, as discussed by Brown and Reilinger. As of this writing, it cannot be said that there are well-confirmed level changes within tectonic plates other than those of obvious local or external cause: water withdrawal, magma motion associated with volcanism, or post-glacial rebound.

Measurements of stress within tectonic plates, discussed by Gay, show more systematic relations to plate motions, both in the sign and orientation of stress. Most intraplate stresses are compressive, perhaps a consequence of the drive for plate motions coming from the margins, rather than by drag along the bottom. Within regions on the order of 1000 km extent the orientations of stresses are generally consistent. However, over greater distances there are variations in some major plates, such as North America, perhaps indicative of some contribution to the stress system from below the plates. Less systematically relatable to stress measurements are surface loadings and unloadings, such as those due to erosion.

The development of improved laser distance measuring methods, radio interferometry, and improved gravimeters is expected to permit better monitoring of long-term distortion in plate interiors.

Of course, the same techniques, plus improved tilt and strain measurement methods, are applicable to geodynamics studies in seismic zones at plate boundaries. In addition, the laser ranging and radio interferometry techniques are likely to lead to the rates of relative motion between the major tectonic plates being measured directly for the first time in the next decade. To avoid duplication in the reports of other ICG Working Groups, the improved geodetic measurement techniques which are expected to be useful for all of these applications are described in the article by Bender.

RELEVELING DATA IN NORTH AMERICA:
IMPLICATIONS FOR VERTICAL MOTIONS OF PLATE INTERIORS

Larry D. Brown and Robert E. Reilinger

Department of Geological Sciences, Cornell University, Ithaca, New York 14853

Introduction

Precise leveling is the most commonly used method for detecting and mapping vertical movements of the earth's surface over periods of days to decades. Compared with currently available alternatives, leveling is more stable over large periods of time and long baselines, yet is less costly and more accurate over short to moderate distances. However, application of leveling measurements to the investigation of vertical tectonics of plate interiors has had mixed results. While studies to date have demonstrated the potential usefulness of leveling, they have also drawn attention to our inadequate understanding of both the accuracy of the measurements used to infer crustal motion as well as the tectonic processes responsible for such motion. Leveling observations of vertical movement in seismic areas, while in many cases poorly understood, are nevertheless often accepted as geologically significant because such movements are not unexpected in tectonically active areas and can be associated with likely causative mechanisms. On the other hand, similar measurements have been interpreted as indicating substantial vertical motion in 'stable' plate interiors, where tectonic activity is generally unexpected, and plausible mechanisms are difficult to identify. The latter group of observations especially has raised some fundamental questions concerning the reliability of leveling estimates of vertical movement (at least in some cases), the time behavior of any true movements, and the nature of the neotectonic forces which may be currently deforming plate interiors. Although apparent movements of the 'stable' interior are more difficult to understand, and are thus subject to correspondingly more skepticism, they are potentially the more informative regarding heretofore unsuspected or unrecognized intraplate neotectonic processes. This report attempts to briefly review and summarize the current status of research using leveling measurements as a guide to intraplate tectonics in North America, identify those related topics where further investigation is most needed, and offer suggestions regarding which approaches are most likely to effectively address the critical issues. We believe that the North American results are representative of similar measurements in other countries, and therefore that the inferences drawn from them are universally applicable.

Although significant and even pioneering work using Canadian leveling results has been reported, by far the most extensive data set of relative (leveling can only give relative changes in height between two points--choice of datum is arbitrary, but often taken to be sea level) elevation change estimates in North America is taken from First- and Second-order measurements collected during the past 100 years by the National Geodetic Survey (NGS; formerly the United States Coast and Geodetic Survey). Virtually all of these measurements were made for geodetic, rather than geodynamic purposes, the respective requirements of which are not necessarily congruent. Those parts of the U.S. Level Net for which more than one leveling, and therefore a vertical motion estimate, are available and shown in Figure 1 (excluding Hawaii). Areas where crustal movements studies have been reported and are discussed here are indicated by the stippled pattern. Sufficient quantities of leveling data in North America have now been examined to justify the general inferences developed in this review.

Vertical Movements in Seismic Areas

Earthquake Movements

The most dramatic vertical motions are those associated with earthquakes. Although the bulk of the world's seismicity is concentrated at plate margins, earthquakes are not uncommon in many intraplate areas and thus investigations of earthquake related deformation are directly relevant to intraplate tectonics. Studies in seismically active areas indicate that ground movements associated with many earthquakes can be thought of as parts of a periodic accumulation and release of strain energy. This sequence, sometimes called the 'earthquake cycle' (Figure

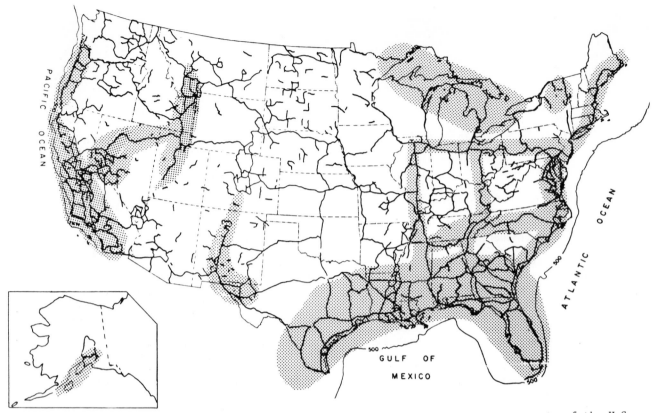

Fig. 1. Lines for which elevation change data are available from leveling measurements of the U.S. Level Net. Stippling denotes regions corresponding to published studies of vertical movements using leveling.

2; Mescherikov, 1968; Scholz, 1972), reflects the long term strain (and stress) buildup between major earthquakes (the interseismic phase), short term precursory deformations immediately before an event (the pre-seismic phase), rapid stress release during the earthquake (the co-seismic phase), and possible relaxation phenomena immediately after the earthquake (the post-seismic phase). Vertical motions corresponding to all four phases have been identified with North American leveling measurements. Co-seismic movements, generally being largest (sometimes several meters in magnitude) and most rapidly accumulated, are the most commonly reported of the four. Most reports of co-seismic movements in North America are associated with plate boundary earthquakes along the west coast [1906-San Francisco, Cal. (Hayford and Baldwin, 1973); 1933-Long Beach, Cal. (Parkin, 1948); 1964-Prince William Sound, Alaska (Small and Wharton, 1969); 1971-Point Mugu, Cal. (Castle et al., 1977)]. Co-seismic movements were also identified from levelings bracketing the 1952 Kern County (Whitten, 1955; Lofgren, 1966) and 1975 Oroville (Savage et al., 1977), California earthquakes, although it is debatable whether or not these should be considered plate margin (in the strictest sense) or intraplate events. Similar move-

ments have been reported for earthquakes located well away from North America's plate margin proper. The co-seismic deformations associated with the 1954 series of earthquakes in west-central Nevada (Whitten, 1957; Reil, 1957) and

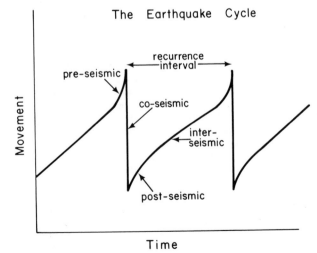

Fig. 2. Relationship between vertical movements of the earthquake cycle.

the 1959 Hebgen Lake, Montana event (Fraser et al., 1964; Myers and Hamilton, 1964) are examples of intraplate tectonic movement successfully monitored by precise leveling. Postseismic movements, identified less often, have been detected by leveling surveys following the 1954 Nevada earthquakes (Savage and Church, 1974), the 1964 Alaskan earthquake (Brown et al., 1978; Prescott and Lisowski, 1977), and the 1971 San Fernando, Cal. earthquake (Savage and Church, 1975). The best documented example of inferred pre-seismic phenomena detected by leveling in North America comes from studies of the 1971 San Fernando event (Castle et al., 1974; 1975). Recently, however, Reilinger (1980) has convincingly argued that most, if not all, of the apparent pre-seismic movements for this event result from leveling errors and non-tectonic subsidence. While the mechanics of earthquakes in all situations is still a matter of investigation, at least some geomechanical models have been proposed to explain the ground movements associated with the various phases of the earthquake cycle: co-seismic deformations in terms of elastic dislocation theory (e.g. Savage and Hastie, 1966), post-seismic in terms of visco-elastic relaxation

phenomena (e.g. Nur and Mavko, 1973), and pre-seismic in terms of dilatancy theories (e.g. Anderson and Whitcomb, 1973). While much of the active research on earthquake deformations is understandably directed toward areas where such events are most common, i.e. at plate boundaries, it is reasonable to expect that similar mechanisms can be associated with earthquakes in intraplate areas as well.

Aseismic Deformation

Vertical movements not explicitly identified with earthquakes, although occurring in seismically active areas, have also been reported in the literature. In most cases the cause of such movement is poorly understood if at all. The most prominent example of these is the so-called 'Palmdale bulge' (Figure 3). This anomalous uplift, reaching a magnitude of over 25 cm relative to reference stations along the coast and affecting an area of over 12,000 square kilometers, was identified by Castle et al. (1976) from leveling measurements in the 'big bend' region of southern California. This movement is unusual in that it is defined by a large number

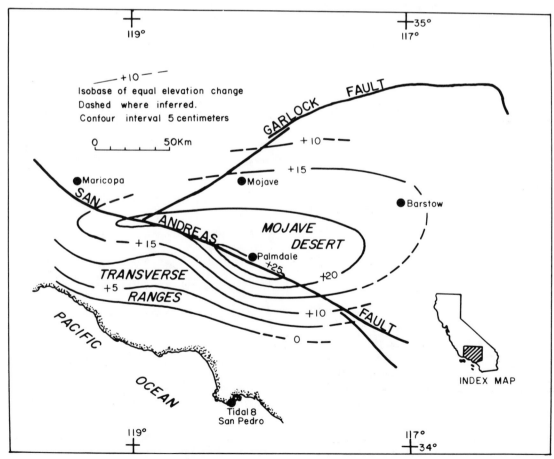

Fig. 3. The Palmdale bulge. From Castle (1978).

of leveling surveys carried out at several different times. Subsequent analyses indicate that the apparent uplift affects a much larger region than initially recognized and, more importantly, that the movement accumulated in a non-uniform manner (Thatcher, 1976; Castle et al., 1977; Bennett, 1977; Castle, 1978). Because of the proximity of the uplift to major faults and cities and the possibility that it could represent a pre-seismic phenomena, it has become the focus of attention for several complementary investigations, including seismic, magnetic, gravity, and geologic studies (see EOS, v. 58, no. 12, 1977). A complete resurvey of the affected area was made in 1978 under the direction of the National Geodetic Survey (EOS, v. 60, p. 50, 1979). This massive project was executed as rapidly as possible to provide a 'snapshot' of the area's geometry in order to avoid complications due to possibly complex time behavior of the movement. Latest results of this work suggest that the bulge has, in fact, deflated (EOS, v. 60, p. 778, 1979).

A completely satisfactory tectonic explanation for the apparent uplift has not been agreed upon, although precursory dilatant uplift (Castle et al., 1975) and compressional strain accumulation (Thatcher, 1976) have been proposed. The debate over the origin of the Palmdale Bulge has re-

cently taken on a radically new complexion with the suggestion that the observed "uplift" is actually the result of systematic leveling errors and near-surface, non-tectonic movements. In particular, topography-correlated errors arising from rod miscalibration and/or atmospheric refraction effects (Jackson and Lee, 1979; Brown et sl., 1980; Jackson et al., 1980; Strange, 1980) are suspected. Although it has not been established to everyone's satisfaction (e.g. Stein and Silverman, 1980) that all of the southern California uplift resulted from misidentification of leveling errors and surficial subsidence, it is clear that these effects are more serious than generally realized and that great caution is required in making tectonic interpretations from releveling observations. If nothing else, the current debate over the reality of the Palmdale Bulge has refocussed needed attention on many of the problems associated with leveling.

Other apparently aseismic crustal movements which do not appear to be related to the San Andreas system have been identified from leveling measurements in other parts of California. Savage et al. (1975) report aseismic (interseismic?) tilting at the base of the Sierra Nevada in Owens Valley. Bennett et al. (1977) and Bennett (1978) also report significant apparent movements in the Sierra Nevada between Roseville,

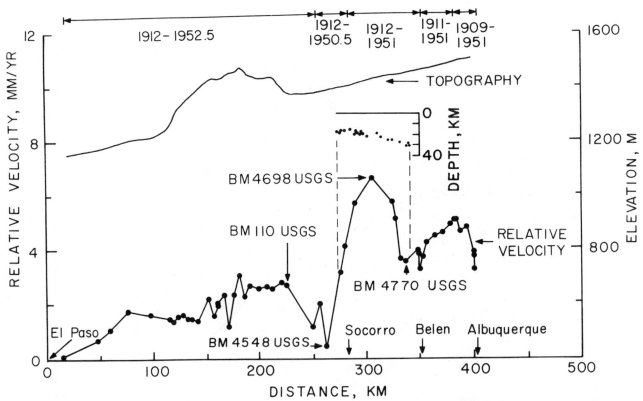

Fig. 4. Recent relative uplift near Socorro, N.M., associated with subsurface magmatic activity. Inset shows top of midcrustal magma body inferred from reflection points for seismic waves. From Reilinger and Oliver (1976).

California and Reno, Nevada, and near Auburn, California, although Chi et al., 1980, have attributed most of the movements along the Roseville-Reno line to systematic error. Whether these should be considered plate boundary or interplate deformations is debatable. The aseismic movements identified by Buchanan-Banks et al. (1975) in the Transverse ranges near Ventura, California are most likely related to strain associated with the San Andreas system.

Vertical Movements and Magmatic Activity

Verical movements associated with volcanic activity have been well known for some time (e.g. Wilson, 1935). Studies of such movements in Hawaii, an intraplate area, are too numerous to describe here: however, most of the observations fit into a rather simple conceptual framework of filling (inflation) and eruption (deflation) of underground magma reservoirs (e.g. Decker, 1969). Of perhaps more interest to the study of interplate phenomena are examples of vertical movements in the continental interior of the U.S. which have been attributed to magmatic activity. The best documented case is reported by Reilinger and Oliver (1976) based on their analysis of leveling measurements in the Rio Grande rift near Socorro, N.M. (Figure 4). The pronounced uplift centered north of Socorro occurs over an area which is believed to be underlain by an extensive mid-crustal magma body. The existence of this chamber has been inferred by Sanford and colleagues from primarily geophysical, and some geological, information (Sanford et al., 1977). The movements reported by Reilinger and Oliver are similar to the deformation computed for models of buried magma bodies consistent with the geophysical studies. Recently releveling near Socorro has confirmed this uplift (Reilinger et al., 1980). Other examples of intraplate movements which may be associated with magmatic activity have been reported for the northern Rio Grande rift (Reilinger and York, 1979), the southern Rio Grande rift in Trans-Pecos Texas (Brown et al., 1978), and near Yellowstone National Park (Reilinger et al., 1977; Smith and Pelton, 1977). These results suggest that leveling can be an effective tool for investigating magmatic processes in intraplate areas and that such activity may be more widespread than heretofore suspected.

Vertical Movements in the 'Stable' Interior

Although much remains to be learned about crustal movements in western North America, that this region should be undergoing such movements is not surprising in view of the many indicators of contemporary tectonic activity, of which seismicity is the most obvious. On the other hand, most of North America east of the Rocky Mountain tectonic belt represents an intraplate area which last experienced major tectonic events

hundreds of millions to billions of years ago. That leveling measurements should indicate significant vertical movements in these areas at the present time, as they do, raises serious questions as to just how 'stable' the interior is, as well as to just how reliable the leveling measurements are as estimates of tectonic activity.

Post-glacial Rebound

Perhaps the only parts of the North American platform-shield complex where one might expect vertical movements sufficiently large to be detected by leveling techniques are those which are likely to be undergoing post-glacial rebound as in Fennoscandia (e.g. Kaariaien, 1953). Unfortunately leveling measurements of elevation change in relevant parts of North America are relatively rare. Results from a small level net in the Lac St. Jean area of Quebec have been interpreted as consistent with post-glacial rebound (Frost and Lilly, 1966; Gale, 1970; Vanicek and Hamilton, 1972). Vanicek (1975, 1976) and Vanicek and Nagy (1980) use scattered leveling results in the Maritime Provinces to infer crustal movements which may result from post-glacial effects (Grant, 1970). Moore (1948) used some leveling in conjunction with water level measurements along the Great Lakes to infer a pattern of broad uplift consistent with that expected from post-glacial rebound. Walcott's (1972) more extensive analysis of lake level data for the Great Lakes also suggests a broad rebound phenomena (Figure 6). Walcott (1972) also interpreted Hicks' (1972) sea level estimates of vertical movement along the east coast of the U.S. as consistent with forebulge collapse following deglaciation. Fairbridge and Newman (1968) reach a similar conclusion with respect to leveling estimates of tilting along the

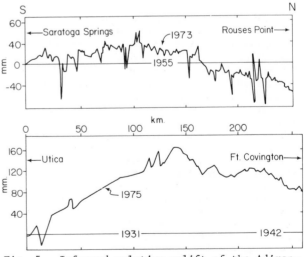

Fig. 5. Inferred relative uplift of the Adirondack dome of N.E. New York. From Isachsen (1976).

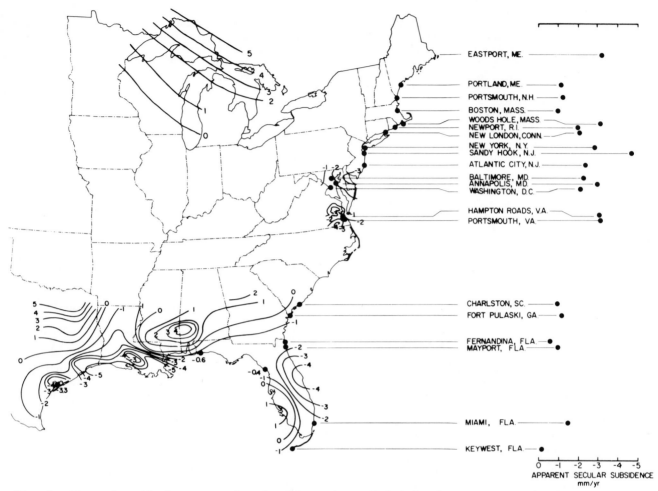

Fig. 6. Maps of vertical movement from leveling and sea (lake) level measurements in the eastern United States. Great Lakes contours by Walcott (1972). Chesapeake embayment and Gulf coast maps by Holdahl and Morrison (1975). Sea level profile by Hicks (1972). From Brown and Oliver (1976).

Hudson River in southeast New York state.

On the other hand, analysis of U.S. leveling results in the central U.S. by Brown and Oliver (1976; Figure 7), in the Adirondacks (Isachsen, 1976; Figure 5), and along the New England coast (Brown, 1978; Figure 8) show little if any evidence of post-glacial rebound. In some cases, the apparent movements appear in the opposite sense. At present, therefore, there are very few leveling measurements available in those parts of North America expected to be undergoing postglacial rebound, and those measurements that are available are sometimes contradictory.

With respect to a related topic, Crittenden (1963) interpreted leveling results across northern Utah as possibly indicating as much as 1 mm/yr of rebound following the removal of Pleistocene Lake Bonneville's water load. However, Walcott's (1970) reanalysis of shoreline data suggests that equilibrium has been obtained, and therefore that the apparent movements cannot be related to rebound. Again, the

paucity of measurements leaves the matter unresolved.

Adirondacks

Based on leveling measurements along the eastern edge of the Adirondacks, Isachsen (1975) proposed that the Adirondack dome is presently going up relative to surrounding regions. Subsequent releveling of a line across the center of the dome provided supporting evidence for this hypothesis (Figure 5, Isachsen, 1976). Correlation of uplift with a major domical geological feature is especially intriguing in view of recurring seismicity at the center of the dome (Sbar and Sykes, 1973). However, lake level measurements appear to contradict those leveling estimates (Barnett and Isachsen, 1980). Furthermore, other level lines in the area, although of poorer quality than those cited by Isachsen, are not fully consistent with the hypothesized doming, thus leaving the interpretation open to some question.

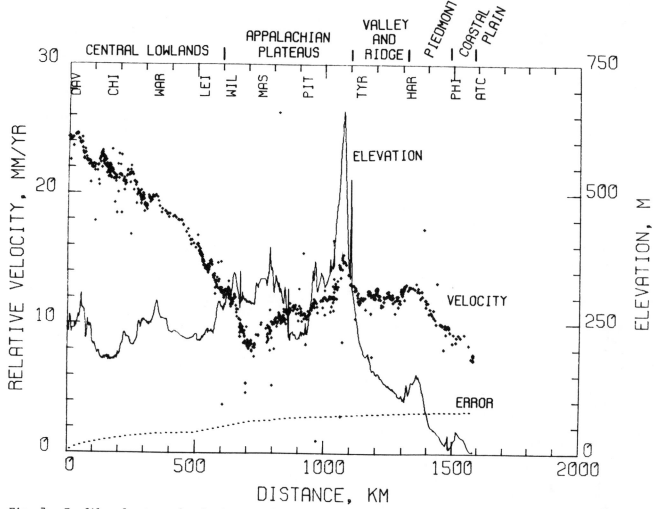

Fig. 7. Profile of rates of relative vertical movement between Davis Junction, Ill. (left) and Atlantic City, N.J. (right). Tectonic provinces crossed are shown at the top. Note the large eastward tilting of the central lowlands. From Brown and Oliver (1976).

Chesapeake Embayment

Holdahl and Morrison (1975) present a contoured map of apparent crustal movement in the area of the Chesapeake embayment based on a relatively large number of leveling lines (Figure 6). The map shows subsidence of this region relative both to sea level and areas to the north and south. The cause of this apparent subsidence is unclear, although Walcott (1972) has argued that it represents forebulge collapse following deglaciation of areas to the north. Vanicek and Christodulidis' (1974) readjustment of the NGS measurements with a different algorithm produced similar results.

Gulf Coast

In the same paper, Holdahl and Morrison (1975) present a map of apparent crustal movements for the much larger Gulf coastal region (Figure 6). To a large degree the map is dominated by subsidence due to water withdrawal and sediment compaction (e.g. in the Houston-Galveston and Mississippi delta regions). However, some very intriguing trends are evident, including the dome-like uplift northeast of the Mississippi delta and the apparent subsidence of the east central Florida coast. The cause of these patterns have not been identified, although Brown and Oliver (1976) suggest that the doming northeast of the delta may be related to sediment loading of the crust beneath the delta.

The Eastern U.S.: A Regional Perspective

Attempts to examine crustal movements in eastern North America as a whole date back to Small's (1963) map of elevation changes along selected level lines in the U.S. Meade (1971) presented a contoured regional map of crustal movement for the eastern U.S. Both maps suffer from having used adjusted, rather than observed, leveling measurements. Adjusted elevation measurements have been corrected to achieve statistical consistency assuming crustal stability. Clearly then, use of these adjusted measurements to estimate crustal motion is likely to result in serious distortion and inaccuracies. Most recent mapping algorithms. such as those used by Vanicek and Christodulidis (1974), Holdahl and Morrison (1975), and Holdahl and Hardy (1977) use observed, rather than adjusted elevation estimates. Jurkowski et al. (1979) have prepared a new map of apparent elevation change in the eastern U.S. using the method of Holdahl and Hardy (1977). Vanicek and Nagy (1980) present a similar map for parts of Canada.

Mizoue (1967) examined leveling estimates of elevation change along selected profiles in the eastern U.S. as part of his comparative study of movements on different continents. The most detailed examination of unadjusted leveling results on a regional scale in the eastern U.S. is that of Brown and Oliver (1976), who discussed long profiles of apparent elevation change based on unadjusted measurements. This study suggested that much of the eastern U.S. was characterized by vertical movements with rates of several mm/yr and that many of the patterns of apparent movement seem to correlate to some degree with major tectonic structures (Figure 7). This correlation is most interesting since many of these structures are believed to have been inactive for millions of years. Furthermore, the trend of movement in some areas is consistent with Phanerozoic trends in the geologic record, although the apparent rates of movement (several

mm/yr) are at least an order of magnitude larger than average rates estimated from geologic evidence. This so-called 'rate paradox' has led to the hypothesis that contemporary movements are episodic or oscillatory with relatively short periods (~10,000 years) in order to avoid the unreasonable prospect of kilometer high relief in platform areas within a few million years (e.g. Boulanger and Magnitsky, 1974; Brown and Oliver, 1976). The difficulties entailed in this hypothesis are outlined later. Clearly, however, if the leveling measurements are even approximately accurate, the continental interior of the U.S. is surprisingly 'active' in a geodynamic sense.

Leveling vs Tide Gauges

In view of some of the ambiguities involved with using leveling in the east, an independent measurement of vertical movement is highly desirable. Brown (1979) compared leveling estimates of apparent elevation change along the east coast with independent estimates of elevation change provided by sea level records (e.g. Hicks, 1972). Figure 8 shows one of the results of that comparison, which indicates a large systematic discrepancy between the two estimates. Uncertainties in tying together the leveling results over different time periods and gaps in data coverage are inadequate to explain the discrepancy. Brown et al. (1979) report a similar discrepancy along the west coast. The most likely, but by no means the only, explanation is that some type of systematic leveling error affects the measurements. Relevant to this discussion is the similar discrepancy between leveling and oceanographic estimates of the sea slope along the coasts (e.g. Sturges, 1968). The west coast

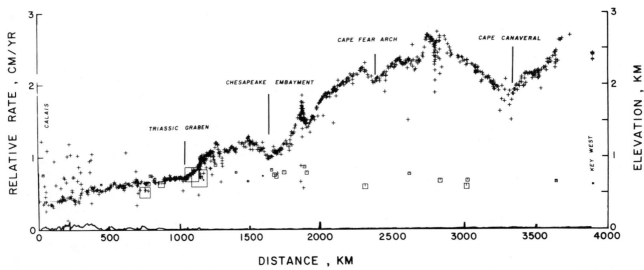

Fig. 8. Comparison of rates of crustal movement indicated by leveling (crosses) and sea level (squares) measurements along the U.S. east coast between Calais, Me. and Key West, Fla. Two estimates set equal at Portland, Me. Note large systematic discrepancy as one goes south. From Brown (1978). Topography, solid line at bottom, is minimal.

discrepancy was recently resolved by new leveling, which now shows agreement with the oceanographic estimates (Balazs and Douglas, 1979). In view of these results and the fact that certain types of error are more likely to accumulate on north-south lines (Bomford, 1971), considerable caution is required in interpreting at least the longer wavelength movement trends. However, as pointed out by Brown (1978), the most significant moderate wavelength 'anomalies' along the east coast profile appear to correspond to major geologic structures, a correlation difficult to account for by appealing to systematic errors or coincidence. Stewart (1975) interprets movements along the coast near the Cape Fear arch as evidence for pre-seismic phenomena, although Brown (1978) suggests other possibilities.

Discussion

Crustal Movement or Systematic Error?

Because the large rates of apparent vertical crustal movement indicated by leveling in the central and eastern United States is seemingly contrary to what would be expected for a 'stable' intraplate region, and because of apparent inconsistencies among leveling results and between leveling and tide gauge measurements, the question naturally arises -- how accurately do leveling measurements over these distances reflect true ground movement as opposed to the systematic accumulation of small measurement errors. For engineering purposes, least-squares adjustment of leveling measurements assuming a static earth are probably quite adequate to maintain closure criteria. However, for crustal movement studies, where utmost accuracy is demanded over long survey distances, it is not clear whether normal precautions to prevent accumulation of systematic effects are effective in all cases. For example, considering the apparent discrepancies along the coasts, it is questionable whether current procedures eliminate certain north-south effects (Reilinger et al., 1980).

Another troublesome type of systematic error in leveling occurs when surveying uphill or downhill. Rod miscalibration or unequal refraction effects can generate errors which correlate with topographic relief (e.g. Kukkamaki, 1938). The result can be an apparent change in elevation which does not correspond to any real ground motion. Citron and Brown(1979) discuss an example in the southern Appalachians of apparent elevation changes which suspiciously correlate with topography (Figure 9). Does this correlation indicate refraction errors or a real association between uplift and elevation, or both? Further complicating the issue is nearby geomorphic evidence which can be interpreted as supporting recent uplift. Similar movement/elevation correlations are numerous in the NGS data base, prompting Brown et al. (1980 b) to argue

that many of the apparent "uplifts" in areas of topographic relief are artifacts of systematic error. Included among such suspect movements would be the Appalachian uplift discussed by Brown and Oliver (1976) and the Adirondack doming found by Isachsen (1976).

Other spurious causes of apparent elevation change which could be mininterpreted as deep-seated neotectonic activity are generally less problematical. For example, Karcz et al. (1976) has questioned the stability of bench marks in the northeastern U.S. While undoubtedly a factor when considering movements over small distances, it is difficult to see how such instability could contribute more than a random error over tens of kilometers or more of leveling. Fluid withdrawal is certainly known to cause massive subsidence in certain areas (e.g. Poland and Davis, 1969) but is unlikely to be mistaken for tectonic activity during a careful investigation. The effect of some near surface processes, such as ground water variations, soil expansion, thermal expansion, etc. have not been thoroughly evaluated. Whitcomb (1976) has reemphasized that leveling measurements are referred to the local geoid rather than a geometrical coordinate system and that gravity variations due to subsurface density changes may be indistinguishable from geometrical elevation changes.

On the other hand there are persuasive arguments against systematic errors having undue influence. If two levelings are carried out under similar circumstances, many systematic effects will tend to influence each leveling equally and therefore cancel out when elevation differences are computed. Many of the most suspect sources fall into this category. In most cases of interest, the measured movements seem to exceed the magnitude of errors expected from these various sources. Co-seismic deformation, for example, commonly exceeds any reasonable systematic error effects by several orders of magnitude. For this and many other types of motion there is virtually no question that real movements are being monitored. Many of the subtler and more difficult to understand intraplate movements, such as those cited above, also exceed commonly cited error limits. The important question is: how realistic are these error limits and how should they be applied?

Adjustment of networks of leveling, if properly carried out, should serve to minimize the contribution of at least some forms of systematic error. When adjusted, elevation differences are corrected in such a way as to eliminate misclosures in leveling circuits. In the past, these adjustments have been made assuming a static earth. New algorithms (e.g. Holdahl and Hardy, 1979) incorporate movement into the adjustment scheme, although it remains to be seen if crustal motion can be satisfactorily modeled with this approach.

In the absence of deterministic evidence regarding the role of systematic errors in many

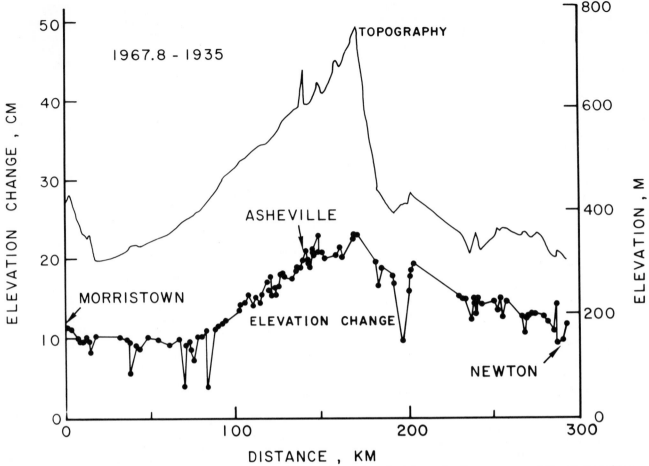

Fig. 9. Relative vertical movement across the southern Appalachians in Tennessee and North Carolina. Note strong correlation between apparent uplift and topography. Tectonics or systematic error? From Citron and Brown (1979).

cases, Brown et al. (1980 a) have suggested some empirical criteria for evaluating the reliability of leveling estimates of crustal movement. These criteria consider a) the magnitude of the apparent movement in excess of known sources of error, b) consistency among all available levelings, c) consistency with independent estimates of movement, d) lack of correlation with topography, e) consistency with reasonable tectonic processes, and f) correlation with geologic and/or tectonic structure. Brown et al. emphasize that these are only factors to be considered when interpreting leveling results, not rigid rules of evidence.

Complex Time Behavior

Interpretation of leveling measurements is further complicated by the fact that true vertical motions in many areas may have a complex time behavior. Earlier treatments have generally assumed that 'secular', slowly varying rates of movement were being measured. As more multiple relevelings become available, it is becoming more apparent that crustal motions may accumulate at

varying rates, even over time periods of a few years. Time dependent rates have been reported for the Palmdale bulge (Thatcher, 1976; Castle, et al., 1977) and in Alaska (Brown et al., 1977), and suspected in other regions (e.g. Brown et al. 1978). Attempting to monitor such behavior with one or two levelings every 25 or 30 years is clearly inadequate and could result in temporal aliasing. Some of the inconsistencies noted in leveling measurements and attributed to systematic errors may, in fact, result from such temporal rate variations.

Over the longer term geodetic measurements must be consistent with geomorphic and geologic estimates of vertical motion. The rapid rates suggested by leveling have been interpreted to require oscillatory or episodic mechanisms to avoid contradictions between these two data sets. However, if such motions are oscillatory or episodic, why should activity be so rapid at the present time? Why should this not be a period of relative quiescence? A satisfactory solution to this dilemma has not been demonstrated, except, of course, for the case of seismic movements. On the other hand, given that horizontal movements with substantially greater rates than those indicated by leveling are accpeted as commonplace (plate motions), rapid and complex vertical motions of plate interiors do not seem so unreasonable. Of course, it is also possible that many of these vertical "movements" may yet prove to be the result of systematic leveling errors, an alternative solution to the "rate paradox".

Cause of Vertical Motions

In many cases a plausible geotectonic mechanism can be invoked to explain leveling observations of vertical motion. This is often the case for movements detected in seismically active regions, and is an important factor in evaluating their reliability. Some degree of understanding has been achieved for earthquake and magmatic deformations. In contrast, the causes of many of the apparent movements indicated by leveling in the 'stable' interior are speculative at best. Brown and Oliver (1976) review a number of proposed mechanisms. Most have difficulty explaining the rapid rates or particular patterns involved. Quantitative modeling of some of the more attractive possibilities, such as strain concentration at reactivated zones of crustal weakness, should be a major goal in future work on contemporary dynamics.

Status of Current Research

The study of vertical motions in intraplate North America is relatively healthy. Studies such as those reviewed here have identified many of the critical issues which need to be resolved before a credible model for contemporary intraplate tectonics can be formulated. Particularly encouraging is the growing recognition among the geodetic community of the need for leveling measurements specifically for geodynamic purposes. Previous efforts in this regard have generally been limited, for example, to surveys at California fault sites (e.g. Holdahl, 1970). The NGS has established a Geodynamics Group, and has begun to undertake surveys at new sites of geodynamic interest to outside scientists (EOS, v. 59, pl 450, 1978). Proposed sites are evaluated by the NGS and the National Research Council of the National Academy of Sciences. The first results from this program have been most encouraging (Reilinger et al., 1980). In addition, the NGS has embarked upon a major re-leveling of the U.S. network, which should provide vast amounts of new data relevant to problems of intraplate dynamics, and has begun a close re-examination of possible source of systematic error. The development of high precision space-oriented techniques for geodetic measurements holds promise for providing complementary, independent crustal motion information, if properly integrated with terrestrial (leveling) results.

Summary

Studies of North American leveling data suggest that, among other things:
a) geologically rapid rates of apparent uplift are surprisingly common, especially in the so-called 'stable' interior
b) many motions can be related to plausible geologic mechanisms, while others cannot
c) the time behavior of some vertical motions may be complex, and multiple relevelings may be necessary to properly monitor them; statistical treatments of the data (e.g. in mapping algorithms) must recognize this possibility.
d) a much greater fraction of precise leveling should be done for geodynamic rather than purely geodetic purposes
e) our present understanding of the role of systematic errors in leveling is inadequate to eliminate serious uncertainties regarding the reliability of some measurements
f) leveling measurements should be integrated with results from geologic, geomorphic, and geophysical investigations to link contemporary motions to longer term processes.

Contemporary crustal motion of plate interiors as revealed by leveling remains a complex, often ambiguous, subject. That intraplate movements of neotectonic origin can be and have been successfully measured by leveling is beyond doubt. That some measurements contain strong suggestions of systematic error which could be misinterpreted as crustal movement also seems well established. The challenge facing geophysicists, geologists, and geodesists alike is to extract reliable tectonic information from measurements which may or may not be contaminated by errors. The practical implications of many of the known or suspected crustal motions (e.g. in seismic hazard evaluation) are so important they demand that the ambiguities connected with the use of leveling to monitor intraplate movements be resolved.

Acknowledgements. The authors wish to acknowledge the advice of Dr. Jack Oliver of Cornell University, and Mr. Sanford Holdahl and Captain John Bossler of the National Geodetic Survey on various aspects of the discussions presented in this paper. The authors were supported by grants from the United States Nuclear Regulatory Commission (AT(49624-0367): Brown) and the United States Geological Survey (14-08-0001-G-110: Reilinger) during compilation of this review.

References

Anderson, D.L., and Whitcomb, J.H., 1973, The dilatancy-diffusion model of earthquake prediction, Proceedings of the Conference on Tectonic Problems of the San Andreas Fault System, Stanford Univ. Publ. Ser. Geol. Sci., v. 13, pp. 417-425.
Balazs, E.I., and B.C. Douglas, 1979, Geodetic

leveling and the sea level slope along the California coast, J. Geophys. Res., v. 84, pp. 6195-6206.

Barnett, S.G., and Isachsen, Y.W., 1980, The application of Lake Champlain water level studies to the investigation of Adirondack and Lake Champlain crustal movements, EΘS, v. 61, p. 210.

Bennett, J., 1977, Palmdale 'bulge' update, California Geology, v. 830, pp. 187-189.

Bennett, J.H., 1978, Crustal movement of the Foothills fault system near Auburn, California, California Geology, v. 31, pp. 177-182.

Bennett, J.H., Taylor, G.C., and Toppozada, T,R,, 1977, Crustal movement in the Northern Sierra Nevada, California Geology, v. 30, pp. 51-57.

Bomford, G., 1971, Geodesy, third ed., England, Clarendon, Oxford, 226 p.

Boulanger, Y.D., and Magnitskiy, V.A., 1974, Contemporary movements of the earth's crust. tate of the problem, Izv. Earth Physics, no. 10, pp. 19-24.

Brown, L.D., 1978, Recent vertical movement along the east coast of the United States, Tectonophysics, v. 44, pp. 205-231.

Brown, L.D., and Oliver, J.E., 1976, Vertical crustal movements from leveling data and their relation to geologic structure in the eastern United States, Rev. Geophys. and Space Physics, v. 14, pp. 13-25.

Brown, L.D., Reilinger, R.E., and Hagstrum, J.T., 1978, Contemporary uplift of the Diablo Plateau, west Texas, from leveling measurements, J. Geophys. Res., v. 82, pp. 3369-3378.

Brown, L.D., Reilinger, R.E., and Citron, C.P., 1980a, Recent vertical crustal movements in the U.S.: evidence from precise leveling, in Earth Rheology and Late GCenozoic Isostatic Movements, John Wiley and Sons, pp. 389-405.

Brown, L.D., D.L. Miesen, R.E. Reilinger, and G.A. Jurowski, 1980b, Geodetic leveling and crustal movement in the U.S.: Part 1, Topography and Vertical Motion, Amer. Geophys. Union Abstract Corr. for 1980 Spring Annual Meeting.

Buchanan-Banks, J.M., Castle, R.O., and Ziony, J.I., 1975, Elevation changes in the central transverse ranges near Ventura, California, Tectonophysics, v. 29, pp. 113-125.

Burford, R.O., Castle, R.O., Church, J.P., Kinoshita, W.T., Kirby, S.H., Ruthven, R.T., and Savage, J.C., 1971, Preliminary measurements of tectonic movement, In: The San Fernando, California Earthquake of February 9, 1971, U.S. Geol. Surv. Prof. Pap. 733, pp. 80-85.

Castle, R.O., 1978, Leveling surveys and the southern California uplift, Earthquake Info. Bull., v. 10, pp. 88-92.

Castle, R.O., Alt, J.N., Savage, J.C., and Balazs, E.I., 1974, Elevation changes preceding the San Fernando earthquake of February 9, 1971, Geology, v. 2, pp. 61-66.

Castle, R.O., Chruch, J.P., and Elliot, M.R.,

1976, Aseismic uplift in southern California, Science, v. 2, pp. 251-253.

Castle, R.O., Church, J.P., Elliot, M.R., and Morrison, N.L., 1975, Vertical crustal movements preceding and accompanying the San Fernando earthquake of February 9, 1971: A summary, Tectonophysics, v. 29, pp. 127-140.

Castle, R.O., Church, J.P. Elliot, M.R., and Savage, J.C., 1977, Pre-seismic and coseismic elevation changes in the epicentral region of the point Mugu earthquake of February 21, 1973, Bull. Seismol. Soc. Amer., v. 67, pp. 219-231.

Castle, R.O., Elliot, M.R., and Wood, S.M., 1977, The southern California uplift, Trans. Am. Geophys. Union, v. 58, p. 495.

Chi, S.C., R.E. Reilinger, L.D. Brown, and J.E. Oliver, 1980, Leveling circuits and crustal movements, Jour. Geophys. Res. 85, pp. 1469-1474.

Citron, G.P., and Brown, L.D., 1979, Recent crustal movements in the Blue Ridge and Piedmont provinces from precise leveling surveys, North Carolina and Georgia, Tectonophysics 52, pp. 223-238.

Crittenden, M.D., 1963, Effective viscosity of the earth derived from isostatic loading of Pleistocene Lake Bonneville, J. Geophys. Res., v. 68, pp. 5517-5530.

Decker, R.W., 1969, Land deformation related to volcanic activity in Hawaii, Amer. Phil. Soc. Yearb. 1969, pp. 295-296.

Douglas, B.C., 1978, Geodetic and steric leveling: together at last?, Trans. Amer. Geophys. Union, v. 59, pp. 1000-1001.

Fairbridge, R.W., and Newman, W.S., 1968, Postglacial crustal subsidence of the New York area, Zetshcr. Geomorphologie, v. 12, pp. 296-317.

Fraser, G.D., Witkind, I.J., and Nelson, W.H., 1964, A geological interpretation of the epicentral area--the dual basin concept, U.S. Geol. Surv. Prof. Paper 435, pp. 99-106.

Frost, N., and Lilly, J.E., 1966, Crustal movement in the Lac St. Jean area, Quebec, Can. Surveyor, v. 20, pp. 292-299.

Gale, L.A., Geodetic observations for the detection of vertical movements, Can. J. Earth. Sci., v. 7, pp. 602-606.

Grant, D.R., 1970, Recent coastal submergence of the Maritime Provinces, Canada, Can. J. Earth Sci., v. 7, pp. 679-689.

Hayford, J.F., and Baldwin, A.L., The earth movements in the California earthquake of 1906, U.S. Coast and Geodetic Surv. Report for 1907, Appendix 3, In: Reports on Geodetic Measurements of Crustal Movement, 1906-71, U.S. Dept. of Commerce, Washington, D.C., pp. 67-104.

Hicks, S.D., 1972, Vertical crustal movements from sea level measurements along the east coast of the United States, J. Geophys. Res., v. 77, pp. 5930-5934.

Holdahl, S.R., 1970, Studies of precise leveling at California fault sites, In: Reports on Measurements of Crustal Movement, 1906-1971,

1973, U.S. Dept. of Commerce, Washington, D.C., 20 p.

Holdahl, S.R., and Hardy, R.L., 1978, Solvability and Multiquadric analysis as applied to investigations of vertical crustal movements, NOAA Technical Memorandum, Rockville, MD.

Holdahl, S.R., and Morrison, N.L., 1975, Regional investigations of vertical crustal movements in the U.S. using precise relevelings and mareograph data, Tectonophysics, v. 23, pp. 373-390.

Isachsen, Y.W., 1976, Possible evidence for contemporary doming of the Adirondack Mountains, New York, and suggested implications for regional tectonics and seismicity, Tectonophysics, v. 29, pp. 169-181.

Isachsen, Y.W., 1976, Contemporary doming of the Adirondack Mountains, New York, Trans. Amer. Geophys. Union, v. 57, pp. 325.

Jackson, D.D., and W.B. Lee, 1979, The Palmdale Bulge - An alternate interpretation, EθS, 60, p. 810.

Jackson, D.D., W.B. Lee, C,C, Liu, P.R. Mullen, and C.C. Chang, 1980, Tectonic motions from leveling data (?), EθS, v. 61, p. 367.

Jurkowski, G.L., L.D. Brown, S. Holdahl, and J.E. Oliver, 1979, Map of apparent vertical crustal movements for the eastern United States, EθS, v. 60, p. 315.

Kaariainen, E., 1953, On the recent uplift of the earth's crust in Finland, Suom. Geodeettisen Laitoksen Julka., v. 42, pp. 1-313.

Karcz, I., Morreale, J., and Porebski, F., 1976, Assesment of benchmark credibility in the study of vertical crustal movements, Tectonophysics, v. 33, pp. T1-T6.

Kukkamaki, T.J., 1938, Uber die nivellistischen Refraktion, Suom. Geodeettisen Laitoksen Julka., v. 25, pp. 1-48.

Lofgren, B.E., 1966, Tectonic movement in the Grapevine Area, Kern County, California, U.S. Geol. Surv. Prof. Pap. 550-B, pp. B6-B11.

Meade, B.K., 1971, Report of the subcommission on recent crustal movements in North America, In: Reports on Geodetic Measurements of Crustal Movement, 1906-1971, U.S. Dept. of Commerce, Washington, D.C., 19 p.

Mescherikov, Y.A., 1968, Recent crustal movements in seismic regions - Geodetic and geomorphic data, Tectonophysics, v. 6, pp. 29-39.

Mizoue, M., 1967, Modes of secular vertical movements of the earth's crust, 1, Bull. Earthquakes Res. Inst. Univ. Tokyo, v. 45, pp. 1019-1090.

Moore, S., 1948, Crustal movement in the Great Lakes area, Bull. Geol. Soc. Amer., v. 50, pp. 697-710.

Myers, W.F., and Hamilton, W., 1964, Deformation accompanying the Hebgen Lake earthquake of August 17, 1959, U.S. Geol. Surv. Prof. Paper 435, pp. 55-98.

Nur, A., and Mavko, G., 1974, Post-seismic viscoelastic rebound, Science, v. 181, pp. 204-6.

Parkin, E.J., 1948, Vertical movement in the Los Angeles region, 1906-1946, Trans. Am. Geophys. Union, vp/ 29, pp. 17-26.

Poland, J.F., and Davis, G.H., 1969, Land subsidence due to withdrawal of fluids, in Review of Engineering Geology II, pp. 187-269.

Prescott, W.H., and Savage, J.C., 1976, Strain accumulation on the San Andreas fault near Palmdale, California, J. Geophys. Res., v. 81, pp. 4901-4908.

Reil, O.E., 1957, Damage to Nevada highways, Bull. Seismol. Soc. Amer., v. 47, pp. 349-362.

Reilinger, R.E., Citron, G.P., and Brown, L.D., 1977, Recent vertical crustal movements from leveling data in southwestern Montana, western Yellowstone National Park, and the Snake River Plain, J. Geophys. Res., v. 82, pp. 5349-5359.

Reilinger, R.E., and Oliver, J.E., 1976, Modern uplift associated with a proposed magma body in the vicinity of Socorro, New Mexico, Geology, v. 4, pp. 583-586.

Reilinger, R.E., and York, J.E., 1978, Relative crustal subsidence from leveling data in seismically active part of the Rio Grande rift, Geology, v. 7, pp. 139-143.

Reilinger, R.E., 1980, Apparent uplift preceding the 1971 San Fernando, California earthquake: Probable artifact of systematic errors in leveling, submitted to Science.

Reilinger, R.E., J. Oliver, L. Brown, A. Sanford, and E. Balazs, 1980, New measurements of crustal doming over the Socorro magma body, Geology, 8, pp. 291-295.

Sanford, A.R., Mott, R.P., Jr., Shuleski, P.J., Rinehart, E.J., Caravella, F.J., Ward, R.M., and Wallace, T.C., Geophysical evidence for a magma body in the crust in the vicinity of Socorro, New Mexico, In: The Earth's Crust, Heacock, J.G., editor, Amer. Geophys. Union Mon. 20, pp. 385-404.

Savage, J.C., and Church, J.P., 1974, Evidence for postearthquake slip in the Fairview Peak, Dixie Valley, and Rainbow Mountain fault areas of Nevada, Bull. Seismol. Soc. Amer., v. 65, pp. 829-834.

Savage, J.C., and Church, J.P., 1975, Evidence for afterslip on the San Fernando fault, Bull. Seismol. Soc. Amer., v.65, pp. 829-834.

Savage, J.C., Church, J.P., and Prescott, W.H., 1975, Geodetic measurement of deformation in Owens Valley, California, Bull. Seismol. Soc. Amer., v. 65, pp. 865-874.

Savage, J.C., and Hastie, L.M., 1966, Surface deformation associated with dip-slip faulting, J. Geophys. Res., v. 71, pp. 4897-4904.

Savage, J.C., Lisowski, M., Prescott, W.H., and Church, J.P., 1977, Geodetic measurements of deformation associated with the Oroville, California earthquake, J. Geophys. Res., v. 82, pp. 1667-1671.

Sbar, M.L., and Sykes, L.R., 1973, Contemporary compressive stress and seismicity in eastern North America: and example of intraplate tectonics, Bull. Geol. Soc. Amer., v. 84, pp. 1861-1882.

Scholz, C.H., 1972, Crustal movement in tectonic areas, _Tectonophysics_, v. 14, pp. 201-217.

Small, J.B., 1963, Interim report on vertical crustal movement in the United States, In: _Reports on Geodetic Measurements of Crustal Movement, 1906-71_, U.S. Dept. of Commerce, Washington, D.C., 14 pp.

Small, J.B., and Wharton, C.C., 1969, Vertical displacements determined by surveys after the Alaska earthquake of March 1964, In: _The Prince William Sound, Alaska, Earthquake of 1964 and Aftershocks_, v. 3, pp. 21-23.

Smith, R.B., and Pelton, J.R., 1977, Crustal uplift and its relationship to seismicity and heat flow at Yellowstone, _Trans. Am. Geophys. Union_, v. 58, pp. 495-496.

Stein, R.S., and S. Silverman, 1980, Random elevation-correlated changes in southern California leveling, _EθS, 61_, p. 367.

Stewart, D.M., 1976, Possible precursors of a major earthquake centered near Wilmington-Southport, North Carolina, _Earthquake Notes_, v. 46, pp. 3-19.

Strange, W.E., 1980, The impact of refraction correction on leveling interpretations in California, _J. Geophys. Res._, in press.

Sturges, W., 1968, Sea surface topography near the Gulf Stream, _Deep Sea Res._, v. 15, pp. 149-156.

Thatcher, W., 1976, Episodic strain accumulation in Southern California, _Science_, v. 194, pp. 691-695.

Vanicek, P., 1975, Vertical crustal movements in Nova Scotia as determined from scattered geodetic relevellings, _Tectonophysics_, v. 29, pp. 183-189.

Vanicek, P., 1976, Patterns of recent vertical crustal movement in Maritime Canada, _Can. J. Earth Sci._, v. 13, pp. 661-667.

Vanicek, P., and Christodulidis, D., 1974, A method for the evaluation of vertical crustal movement from scattered geodetic relevellings, _Can. J. Earth Sci._, v. 11, pp. 605-610.

Vanicek, P., and Hamilton, A.C., 1972, Further analysis of vertical crustal movement observations in the Lac St. Jean area, Quebec, _Can. J. Earth Sci._, v. 9, pp. 1139-1147.

Vanicek, P., and D. Nagy, 1980, The map of contemporary vertical crustal movements in Canada, _EθS_, 61, p. 145.

Walcott, R.I., 1972, Late Quaternary vertical movements in eastern North America: Quantitative evidence of glacial-isostatic rebound, _Rev. Geophys. Space Phys._, v. 10, pp. 849-884.

Whitcomb, J.H., 1976, New vertical geodesy, _J. Geophys. Res._, v. 81, pp. 4937-4944.

Whitten, C.A., 1955, Measurements of earth movements in California, _California Dept. Nat. Resources, Div. Mines Bull._, v. 171, pp. 75-80.

Whitten, C.A., 1957, The Dixie Valley-Fairview Peak, Nevada, earthquakes of December 16, 1954; geodetic measurements, _Bull. Seismol. Soc. Amer._, v. 47, pp. 321-325.

Wilson, R.M., 1935, Ground surface movement at Kilauea volcano, Hawaii, _Univ. Hawaii Res. Publ._, v. 10, pp. 16-47.

THE STATE OF STRESS IN THE PLATES

N.C. Gay

Bernard Price Institute of Geophysical Research
University of the Witwatersrand, Johannesburg

Most of our knowledge on the state of stress within the plates that form the crust of the earth comes from measurements made for rock mechanics purposes in deep mines and near surface engineering projects, from hydraulic fracturing of boreholes and from earthquake source studies. These observations show that, in general, the state of stress at a point within a plate is not simply that due to the weight of the overlying rocks but is the resultant of several preexisting and present day force fields (Figure 1). Of these components some, such as those induced by man, may only have local influence in the immediate vicinity of an excavation, while others, due to relatively small geologic structures, may affect the state of stress within a large mine. By contrast, forces involved with active or pre-existing, global or continental tectonics should result in a relatively uniform state of stress over large areas of a plate.

This is not a new observation. In 1942, E. M. Anderson concluded from his study of faulting in the British Isles that "the stresses which gave rise to it were anything but local in character" and that "at any given period...the same general system of stress...affected great parts of the country." Present-day knowledge of the state of stress in the earth's crust is sufficient to confirm Anderson's observation and perhaps show that uniform systems of stress can operate over large parts of continents. In particular, the orientations of the ambient principal stresses can be fairly well defined. Estimates of absolute magnitudes are more conjectural and need to be improved if we are to add to our understanding of geodynamic processes, such as the driving mechanism for plate motion or even the magnitude of shear stresses across major tectonic fault zones.

Most instruments used for determining in-situ stresses in rock are based on the measurement of strain; i.e. a change in strain of the rock is recorded, from which the change in stress can be deduced. The process involves inserting a device capable of monitoring deformation down a borehole, relieving the stresses acting on the rock around the borehole by overcoring and recording the accompanying change in strain. If the elastic moduli are known, the strain changes can be correlated with the relaxed stresses. Stress-measuring instruments commonly in use include the U.S.B.M. borehole deformation cell, the C.S.I.R. "doorstopper" and triaxial strain cells and Hast's [1958] stressmeter. The use of these and other instruments is discussed in several review articles, e.g. Leeman [1969]; Fairhurst [1968]; McGarr and Gay [1978].

Hydraulic fracturing, or the "hydrofrac technique," is probably the only method that enables measurements to be made at a large distance from a free surface. The technique, which is discussed in detail by Haimson and Fairhurst [1970] and Zoback et al. [1977], involves pumping fluid under pressure into an isolated portion of a well or a borehole until a tensile fracture is induced in the borehole. The breakdown and shut in pressures and the orientation of the induced fracture are used to determine the orientation and magnitudes of the principal stresses although, usually, one of the principal stresses is assumed to be vertical and is calculated from the weight of the overburden. There is some controversy about the interpretation of hydrofrac data, particularly as regards the determination of the maximum horizontal principal stress [cf. McGarr and Gay, 1978].

Earthquake source studies such as fault plane solutions of seismic radiation patterns and the determination of stress drops from seismic moments provide information on the orientation of principal stresses and the magnitude of stress changes accompanying earthquakes. These techniques have the advantage of providing information on stress orientations at substantial depths within the plates, as against the near-surface techniques discussed above, but apart from estimates of stress drops [0.1-10 MPa; Hanks, 1977] no information about stress magnitudes is obtained.

Measurements of in-situ stress, using either the overcoring or hydrofrac methods, are available from all continents, except South America and Antarctica (to my knowledge). Where the complete stress tensor has been determined, it appears that the principal stresses are not

Fig. 1. Components of natural and man-induced stresses that can contribute to the present day in-situ stress at a point in the earth's crust.

normally oriented vertically and horizontally. For example, in Southern Africa (Figure 2) many of the principal stresses cluster within about 30° of the vertical and others tend to plot near the horizontal but departures from these directions are common. The magnitudes of the near vertical principal stresses are generally within twenty percent of the overburden stress. In fact, there is a very good linear correlation between the measured vertical stress and the weight of overburden (Figure 3) and marked deviations from this rule can generally be attributed to local geologic structures.

The variations in magnitudes of horizontal stress are considerably greater than for the vertical stress and although there is a regular increase in the stresses with depth there is no simple linear correlation. There also appear to be significant magnitude differences among the various regions for which data are available

(Figure 4). Thus, in Southern Africa, the average horizontal stress is generally greater than the vertical stress at depths of less than 500 m but becomes smaller than it at greater depths. By contrast, in Canada, the average horizontal stress is larger than the vertical stress at all depths for which data are available. Typically, the horizontal stresses in Canadian mines are at least twice as large as horizontal stresses at equivalent depths in the Witwatersrand gold mines of Southern Africa. The data from the United States of America (determined mainly by hydraulic fracturing) tend to fall between the extremes of Southern Africa and Canada and they also display much less scatter. Only values of the minimum horizontal stress are plotted because of the possible uncertainty in the maximum magnitudes. As noted by McGarr and Gay [1978], the near surface data (<2.3 km depth) tend to increase linearly with a gradient of 15 MPa/km but, at greater depths, the rate of increase in minimum σ_H is faster. Generally the magnitudes of the minimum stresses are less than the vertical stresses except very near surface.

Available data from Europe and the U.S.S.R. show a more complex pattern. In Scandinavia, two trends can be discerned: in the basement gneiss terrain the horizontal stresses are significantly larger than the vertical stresses while in the Caledonides, the vertical stress generally exceeds the horizontal stress. The Russian data fall between these two extremes.

The variation of horizontal stress with depth in Australia shows a relatively wide scatter and

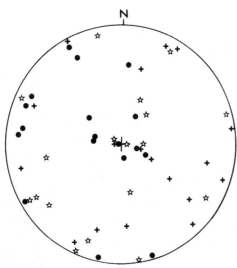

Fig. 2. Lower hemisphere, equal area projection of the directions of principal stress at sites in Southern Africa [after Gay, 1975,1977; Van Heerden, 1976]. Circles - maximum principal stress σ_1; crosses - intermediate principal stress σ_2; stars - minimum principal stress σ_3.

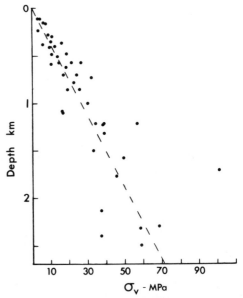

Fig. 3. Magnitudes of vertical stress at depths greater than 100 m. The dashed line corresponds to the theoretical overburden stress, assuming a density of 2700 kg/m³ [After McGarr and Gay, 1978.]

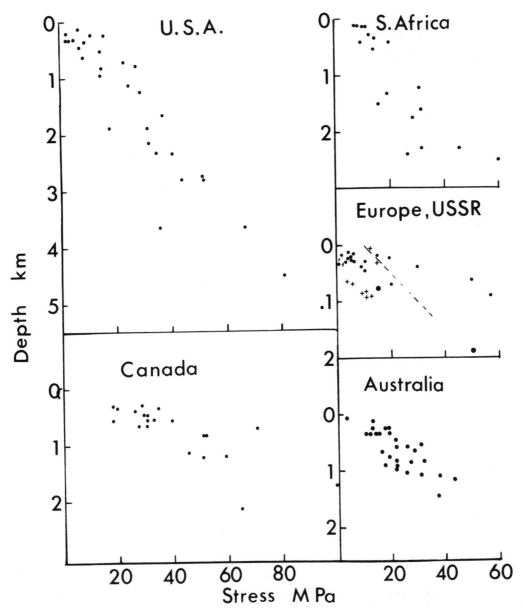

Fig. 4. Magnitudes of horizontal stress. In these graphs, the <u>average</u> horizontal stress is plotted except for measurements made by hydraulic fracturing, in which case the minimum value is used. Sources of data: <u>U.S.A.</u> - McGarr and Gay [1978]. <u>Canada</u> - McGarr and Gay [1978], Sanden [1971]. <u>S. Africa</u> - Gay [1975,1977], Van Heerden [1976]. <u>Europe, U.S.S.R.</u> - Brückl <u>et al.</u> [1975], Bulin [1971], Hast [1958,1969,1973], Li [1971]. The dashed line is the best fit to Hast's Scandinavian basement measurements; crosses - Scandinavian Caledonides; small dots - U.S.S.R.; large dots - Alps. <u>Australia</u> - Worotnicki and Denham [1976].

generally these stress components are greater in magnitude than the vertical stress.

The differences in variation of vertical and horizontal stress with depth, in the regions for which data are available, are well displayed by graphs (Figure 5) showing the change in the ratio σ_H/σ_V with depth, where σ_H is the average horizontal stress except for the hydrofrac data where it is the minimum horizontal stress. The curves

for Southern Africa and North America show well defined trends with the ratio having an approximately constant value at depths below a kilometre and increasing to large values greater than unity near surface. The approximate average values of σ_H/σ_V at depth are 0.6 for Southern Africa, 0.7 for the United States of America and 1.2 for Canada.

The picture for the European and Russian data

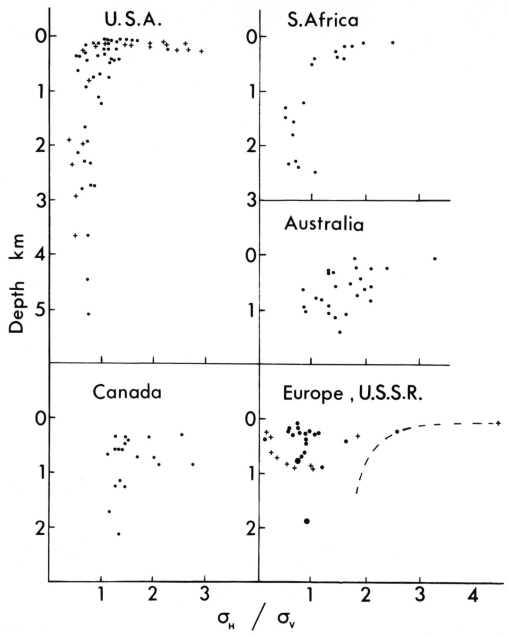

Fig. 5. Variation with depth of the ratio σ_H/σ_V. The <u>average</u> horizontal stress is used to compute the ratio except for hydraulic fracturing measurements, in which case the minimum stress is used. Sources of data as for Figure 4 plus, for U.S.A., H. Swolffs [unpublished data] and, for Europe, Cook [1976]. In the U.S.A. plot, dots – measurements made in relatively soft sediments; crosses – measurements in hard rocks (granites, gneisses, quartzites). In the plot for Europe, U.S.S.R., the dashed curve is Cook's [1976] interpretation of the Scandinavian basement data.

is more complicated. The data from the Scandinavian basement terrain follow a similar curve to those described above but the value of the ratio is very much higher, tending to 1.8 at depth. By contrast, σ_H/σ_V for the Caledonide and Alpine data decreases in value from about 1 to 0.1 with <u>decreasing</u> depth while the Russian data have an approximately constant ratio at depth and both

increasing and decreasing values near surface. Finally, the Australian data show no trend; there is a wide scatter of points and values of σ_H/σ_V range from 0.9-2.5 over the entire depth range.

McGarr and Gay [1978] plotted the variation in maximum shear stress with increasing depth for sites in the United States, Canada, Australia and Southern Africa. This plot showed a very wide scatter in data points but there was a general increase in shear stress with depth, with the rate of increase more rapid in the upper levels. They also noted that shear stresses in "hard rocks" (granite, quartzite, norites) tended to be larger than those in "soft rocks" (sandstones, limestones, shales) and that, possibly, there was a lower limit of 20–40 MPa for shear stresses at mid-crustal depths. Perhaps a better way of estimating the available energy for deformation within the crust is to compute the octahedral shear stress $\left(\tau_{oct} = 1/3\left[(\sigma_1-\sigma_2)^2+(\sigma_2-\sigma_3)^2+(\sigma_3-\sigma_1)^2\right]^{1/2}\right)$ which allows for the influence of all the principal stresses. Figure 6 is a plot of the variation of this parameter with depth for data from the United States of America and Canada. It shows two well-defined trends. The United States of America data cover the greater depth range and show a rapid increase in τ_{oct} until a depth of about 2 km, after which the rate of increase is very much less, possibly zero. By contrast, the values of τ_{oct} in Canadian mines are significantly larger and increase with depth at a much more rapid rate.

In summary, the available information about stress in the upper parts of the crust allows us to make the following observations about stresses within plates: The principal stress components of the stress tensor are normally unequal in magnitude and are inclined to the vertical and horizontal. The magnitudes of vertically oriented stresses are approximately equal to the theoretical overburden stress due to the weight of the strata overlying the point of measurement. Horizontal stresses vary greatly in magnitude at a given depth but they generally exceed the vertical stress at shallow depths and may become smaller than it at greater depths. This change is

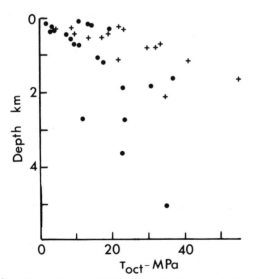

Fig. 6. Variation with depth of the octahedral shear stress (τ_{oct}) at measuring sites in Canada (crosses) and the United States of America (dots). Sources of data as for Figure 4.

well displayed by the variation in the ratio of average horizontal stress to vertical stress. In Southern Africa and the United States of America, this ratio tends to have a constant value of less than one at depths greater than one kilometre while in Canada it tends to be greater than one. At shallower depths, the ratio increases as the horizontal stress becomes 2–3 times the vertical stress very near ground surface. In Europe, particularly Scandinavia, there appear to be two trends: in the Scandinavian Caledonides, σ_H/σ_V tends to decrease with decreasing depth while in the basement gneiss terrain it is large and increases with decreasing depth. The Russian data suggest that the ratio is constant with increasing depth and there is no trend discernible in the Australian data. Shear stresses, and particularly the octahedral shear stresses, tend to increase with depth but the rate of increase decreases with increasing depth and perhaps approaches zero.

How is one to interpret these observations? Nearly all theories on the state of stress assume that one principal stress is vertical and has magnitude $\sigma_V = \rho g Z$, where ρ is the average density of the overburden, g is the acceleration due to gravity and Z is the depth. Clearly, this is a valid assumption; although a principal stress is only rarely oriented vertically, it generally tends to be within 30° of the vertical and has a magnitude close to the overburden stress. Theories which assume that the state of stress in the crust is due solely to gravitational loading include Heim's rule and the lateral constraint of an elastic medium. Heim's rule supposes that the stresses at depth tend to become lithostatic (i.e. equal in all directions) because of creep over a long time and have magnitudes $\sigma_1 = \sigma_2 = \sigma_3 = \rho g Z$. Lateral constraint of an elastic medium results in a vertical maximum principal stress of magnitude $\sigma_1 = \rho g Z$ and equal horizontal principal stresses, σ_2, σ_3, with magnitudes $[\nu/(1-\nu)]\rho g Z$, where ν is Poisson's ratio, typically about 0.25. Thus the horizontal stress will always be smaller than the vertical stress.

Neither of these two simple theories are supported by the observed variations in horizontal stress and this is obviously because the assumptions are not realistic. Complicating factors include the depositional, erosional and thermal history of the rocks and the present and pre-existing tectonic conditions.

Models allowing for stress changes during deposition [e.g. Seagar, 1964; Voight, 1966a; Price, 1974] assume that the stresses are derived only from the weight of the overburden, that no horizontal strain is possible and that yielding or creep may occur. During the pre-diagenetic build up of stresses due to the addition of sediment, σ_H/σ_V increases to about 0.7 and, if erosion occurs, a marked increase in the horizontal stress takes place, resulting in values of σ_H/σ_V as high as 3:1. These high ratios could be preserved during diagenesis and lithification, even allowing for an increase in σ_V because of the increased density of the rock.

Changes in the state of stress during reduction of the overburden pressure by erosion under conditions of lateral restraint [Seagar, 1964; Voight, 1966b] also result in an increase in the horizontal stress relative to the vertical stress. Assuming elastic recovery during unloading, a change in depth of $-\Delta Z$ results in reductions of $-\Delta\sigma_V = -\rho g\Delta Z$ and $-\Delta\sigma_H = -[\nu/(1-\nu)]\,\Delta\sigma_V$ in the vertical and horizontal stresses respectively. Thus the stresses vary linearly with decreasing depth but the change in σ_H is proportionately less than that in σ_V. Gay [1975] showed that this model fitted the observed variation of σ_H/σ_V with depth in Southern Africa very well indeed.

Nevertheless, the model is unrealistic because it ignores stresses due to uplift on a sphere and to thermal changes. During uplift on a sphere there is an increase in the horizontal surface area and, as a result, horizontal strains occur. This causes a horizontal tensile stress to be superimposed on the compressive stress induced by gravitational unloading [Price, 1966, 1974; Haxby and Turcotte, 1976]. This tensile stress may be as large as half the change in the vertical stress [Price, 1966] so that the resultant ratio of horizontal to vertical stresses is given by the equation

$$\sigma_H/\sigma_V = \lfloor (\sigma_H/\sigma_V)_i - [(1+\nu)/2(1-\nu)]\Delta\sigma_V/\sigma_{Vi}\rfloor/$$
$$[1-\Delta\sigma_V/\sigma_{Vi}] \qquad (1)$$

where the subscript $\underline{i}$ refers to the initial state.

This equation has some surprising implications because it predicts (Figure 7) that, depending on its initial value, the ratio σ_H/σ_V either decreases, remains constant or increases during uplift. The diverging trends in σ_H/σ_V in the Scandinavian Caledonides and basement gneiss terrain (Figure 4) may be explained, in part at least, in terms of this model.

The thermal stress generated by a temperature

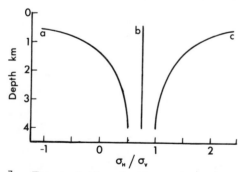

Fig. 7. The variation in the ratio σ_H/σ_V with decreasing depth during uplift, as calculated from equation (1). The assumed initial values of σ_H/σ_V at a depth of 4 km are 0.5, 0.75 and 1.0 for curves a, b, c respectively.

change, ΔT is given by $\sigma_{th} = \alpha E \Delta T/(1-\nu)$ where α is the linear coefficient of thermal expansion and E is the Young's Modulus [Haxby and Turcotte, 1976]. Thus, for an element of rock subjected to bilateral constraint, an increase in temperature induces a compressional stress and a decrease in temperature induces a tensile stress. For a geothermal gradient of 25°C/km, the thermal stresses induced during removal of 1 km of overburden are tensile and of the order of -15 to -30 MPa. These stresses are much larger than those due to erosion and uplift and so, one should expect tensile stresses of about -10 MPa in the upper parts of the crust.

Clearly, this conclusion is in marked contrast with the observed compressive states of stress. However, as pointed out by Voight and St. Pierre [1974], large tensile stresses can only exist if the rock strength is not exceeded. The tensile strength of many rocks is less than 10 MPa and thus fractures or joints should form in the rock on a macroscopic and microscopic scale as a result of the induced thermal stresses. If finite displacements do not accompany this fracturing, the elastic strain energy would not be fully released and lateral compressive stresses of magnitudes between 10-100 MPa would be imposed upon the rock mass. Voight and St. Pierre [1974] point out that horizontal stresses of these magnitudes are known in granites from the Appalachians, where maximum compression directions are approximately perpendicular to the predominant trends of microfractures in quartz.

Another inelastic process which could lead to the enhancement of horizontal stresses is natural hydraulic fracturing of basin sediments [Price, 1974]. This should occur when the pore fluid pressure exceeds the minimum horizontal stress by an amount greater than the tensile strength of the rock. The result is a reduction in pore pressure, and an increase in horizontal stress.

Residual components of palaeotectonic and palaeotopographic stresses can also affect the present day state of stress at individual localities. The influence of these components cannot easily be incorporated in any theoretical model but they have been invoked to explain anomalous stress states in various mines [e.g. Eisbacher and Bielenstein, 1971; Gay, 1975; Herget et al. 1975].

Also the models discussed above do not allow for the varying geological structure in the different parts of the plates. For example, the average horizontal stress tends to be smaller than the vertical stress in sedimentary cover rocks such as in the Witwatersrand basin in Southern Africa, and greater in Precambrian basement terrains, such as in Canada and Scandinavia, and in Palaeozoic fold belts, such as the Urals [Bulin, 1971; Ranalli and Chandler, 1975]. These generalisations probably only hold true in the upper one or two kilometres of the crust; the deep hydraulic fracturing data in the United States (Figure 5) shows that the minimum horizontal

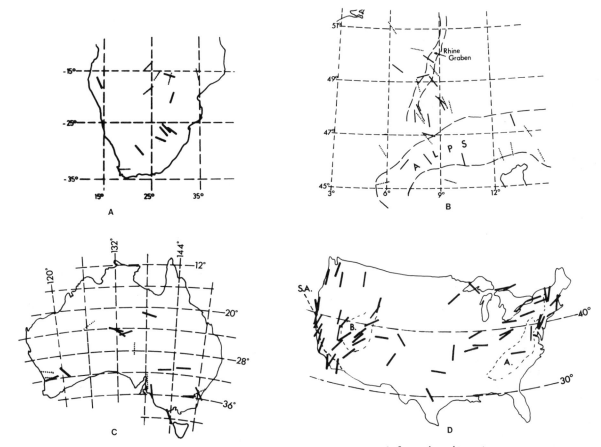

Fig. 8. Orientations of maximum horizontal stresses as determined from in-situ stress measurements (solid lines) and earthquake focal mechanism solutions (dotted lines). Sources of data: Southern Africa – Fairhead and Henderson [1977]; Gay [1977]. The in-situ stress data are the geometric mean orientations calculated from the maximum horizontal stress orientations listed by Gay [1977]. Australia – Worotnicki and Denham [1976]. Central Europe – Ahorner [1975], Greiner and Illies [1977]. United States of America – McGarr and Gay [1978]: A – Appalachians; B – Basin and Range Province; S.A. – San Andreas Fault; no distinction between in-situ stress solutions and focal mechanism solutions.

stresses in Precambrian crystalline rocks are smaller than the equivalent stresses in soft sedimentary rocks at the same depth.

None of the data discussed thus far allows us to confirm or deny Anderson's [1942] observation that the same general stress system may operate over large areas of a plate. To do this we have to look at the orientations of the horizontal stresses. Most of the available data indicate that the horizontal stresses at a measuring site are unlikely to be equal in magnitude and that, in general, there exists a horizontal stress anisotropy [Ranalli and Chandler, 1975] which does not disappear with increasing depth. Thus there may be grounds for correlating the orientations of horizontal stresses determined from shallow stress measurements with stress orientations in the deeper parts of the plates. Such a correlation is supported by the generally good agreement between stress orientations determined from

earthquake source studies and in-situ stress measurements [Ahorner, 1975; Sykes and Sbar, 1974; Worotnicki and Denham, 1976]. Thus maps of horizontal stress orientations should provide information on stress trajectories within plates.

Examples of such maps are shown in Figure 8 for selected areas of the African, Indian, Eurasian and American plates. In Southern Africa, the average maximum horizontal principal stresses tend to be oriented E-W to NW-SE but, in Zambia and areas further north, fault plane solutions indicate that the maximum stress strikes NNE. In Australia most measurements show the maximum horizontal stress acting in an east-west direction, except in the southern-central region where it is oriented approximately N-S.

An especially significant region is that of Central Europe where there is good agreement between stress orientations deduced from shallow stress measurements and crustal earthquakes, most of which indicate that the maximum horizontal stress acts in a NW direction. The data have been interpreted by Illies and Greiner [1976] as indicating an anticlockwise rotation of the maximum stress through about 60° since early Miocene times, when the Rhine graben rift system opened up parallel to the maximum horizontal stress. During the Alpine orogeny, the stress system was reoriented to convert the graben into a left lateral shear zone across which slip is occurring at 0.05 mm per year. The present day stresses are largest in the Central Alps and least in the rift zone, suggesting that very large components of horizontal stress are generated by active plate tectonic forces in the Alpine region.

The data from North America suggest that at least two different stress regimes exist. In the Western United States, the maximum horizontal stress is oriented N to NNE and is consistent with a right lateral shear movement along the San Andreas Fault, part of the boundary between the North American and Pacific plates. Away from the plate margin, the maximum horizontal stress changes in orientation to NE in the Basin and Range Province. Further east, from the middle of the continent to the Appalachians, a similar direction, E to NE, is evident but east of the Appalachians the orientation changes again to N-S.

Recently, Bell and Gough [1979] have used information on oil well break-outs to determine variations in the orientation of maximum horizontal stress in Alberta, Canada. The break-outs occur so as to elongate the boreholes in a northwest-southeast direction, from which they deduce a remarkably uniform orientation for the maximum horizontal stress of northeast-southwest, over a large part of Alberta. They suggest that the uniform stress may result from present tractions acting on the edges and underside of the North American plate. Alternatively, it may represent a residual stress associated with the folding of the Rocky Mountains during the Laramide orogeny.

The significance of maps such as those in Figure 8 is that they demonstrate that, over large parts of the interior of a plate, a system of homogeneous stress orientations may exist to relatively deep crustal levels. This information, coupled with the observation that the shear stress tends to increase with increasing depth, possibly reaching a maximum value, means that deviatoric stresses, capable of inducing rock deformation or fracture over a wide area, exist within plates. These stresses presumably result from active or current tectonic forces generated within the plates during plate motions. Thus, knowledge of the distribution of stresses within plates should provide a check on plate tectonic driving force models.

Acknowledgments. I am grateful to Jay Barton, Art McGarr and Steve Spottiswoode for reviewing this paper. Hans Swolffs generously made some of his unpublished material available.

References

Ahorner, L., Present-day stress field and seismo-tectonic block movements along major fault zones in central Europe, Tectonophysics, 29, 233-249, 1975.

Anderson, E.M., The dynamics of faulting, Oliver and Boyd, Edinburgh, 1942,

Bell, J.S. and D.I. Gough, Northeast-southwest compressive stress in Alberta: Evidence from oil wells, Earth Plan. Sci. Letters, 45, 475-482, 1979.

Brückl, E., K.H. Roch and A.E. Scheidegger, Signinificance of stress measurements in the Hochkönig massif in Austria, Tectonophysics, 29, 315-322, 1975.

Bulin, N.K., The present stress field in the upper parts of the crust, Geotectonics, 3, 133-139, 1971.

Cook, N.G.W., Methods of acquiring and utilizing geotechnical data for the design and construction of workings in rock. In Bieniawski, Z.T. (ed.). Exploration for rock engineering, 1; A.A. Balkema, Cape Town, 1-14, 1976.

Eisbacher, G.H. and H.U. Beilenstein, Elastic strain recovery in Proterozoic rocks near Elliott Lake, Ontario, J. Geophys. Res., 76, 2012-2021, 1971.

Fairhead, J.D. and N.B. Henderson, The seismicity of southern Africa and incipient rifting, Tectonophysics, 41, T19-T26, 1977.

Fairhurst, C., Methods of determining in-situ rock stress at great depths, Tech. Rept. No. 1-68, U.S. Army Corps of Engineers, Missouri River Division, Omaha, Nebraska, 68102, 1968.

Gay, N.C., In-situ stress measurements in Southern Africa, Tectonophysics, 29, 447-459, 1975.

Gay, N.C., Principal horizontal stresses in Southern Africa, Pure Appl. Geophys., 115, 3-10, 1977.

Greiner, G. and J.H. Illies, Central Europe: active or residual tectonic stresses, Pure Appl. Geophys., 115, 11-26, 1977.

Haimson, B.C. and C. Fairhurst, In situ stress determination at great depth by means of hydraulic fracturing, in Rock Mechanics - Theory and Practice, Proc. 11th Symposium on Rock Mechanics, edited by W. Somerton, AIME, 559-584, 1970.

Hanks, T.C., Earthquake stress drops, ambient tectonic stresses and stresses that drive plate motions, Pure Appl. Geophys., 115, 441-458, 1977.

Hast, N., The measurements of rock pressure in mines. Sveriges Geologiska undersökning. Avhandlinger och Uppsater, Ser. C., 52, 1-183, 1958.

Hast, N., The state of stress in the upper parts of the earth's crust, Tectonophysics, 8, 169-211, 1969.

Hast, N., Global measurements of absolute stress, Phil. Trans. R. Soc. Lond. A, 274, 409-419, 1973.

Haxby, W.F. and D.L. Turcotte, Stresses induced by the addition or removal of overburden and associated thermal effects, Geology, 4, 181-184, 1976.

Herget, G., A. Pahl and P. Oliver, Ground stresses below 3000 feet. Proc. 10th Canadian Rock Mechanics Symposium, Queens University, Kingston, 1, 281-307, 1975.

Illies, J.H. and G. Greiner, Regionales stressfeld und Neotektonik in Mitteleuropa. Oberrhein. geol. Abh., 25, 1-40, 1976.

Leeman, E.R., The "doorstopper" and triaxial rock stress measuring instruments developed by the C.S.I.R., J.S. Afr. Inst. Min. Metall., 69, 305-339, 1969.

Li, B., Natural stress values obtained in different parts of the Fennoscandian rock masses. Proc. 2nd Congr. Int. Soc. Rock Mech., Belgrade, 1970, 1, 209-212, 1971.

McGarr, A. and N.C. Gay, State of stress in the earth's crust, Ann. Rev. of Earth Planet. Sci., 6, 405-436, 1978.

Price, N.J., Fault and joint development in brittle and semi-brittle rock. Pergamon, London, 176 pp., 1966.

Price, N.J., The development of stress systems and fracture patterns in undeformed sediments. Proc. 3rd Congress Int. Soc. Rock Mechanics, 1A, 487-496, 1974.

Ranalli, G. and T.E. Chandler, The stress field in the upper crust as determined from in-situ measurements. Geol. Rundschau, 64, 653-674, 1975.

Sanden, B.H., In situ deep mine stress field determinations using a borehole deformation overcoring technique. B.Sc. Thesis, Mining Engineering Dept., Queen's Univ., Kingston, Canada, 1971.

Seagar, J.S., Pre-mining lateral pressures. Int. J. Rock Mech. Min. Sci., 1, 413-419, 1964.

Sykes, L.R. and M.L. Sbar, Focal mechanism solutions of intraplate earthquakes and stresses in the lithosphere. In Geodynamics of Iceland and the North Atlantic Area, edited by L. Kristjansson, D. Reidel Publishing Company, Dordrecht, Holland, 207-224, 1974.

Van Heerden, W.L., Practical application of the C.S.I.R. triaxial strain cell for rock stress measurements. In Exploration for roch engineering, 1, edited by Z.T. Bieniawski, A.A. Balkema, Cape Town, 189-194, 1976.

Voight, B., Interpretation of in-situ stress measurements, Proc. 1st Congress Int. Soc. Rock Mech., Lisbon, 1966, III, 332-348, 1966a.

Voight, B., Beziehung zwichen grossen horizontalen Spannungen in Gebirge und der Tektonik und der Abtragung, Proc. 1st Congress Int. Soc. Rock Mech., Lisbon, II, 51-56, 1966b.

Voight, B. and B.H.P. St. Pierre, Stress history and rock stress. In: Advances in rock mechanics. Proc. 3rd Congress Int. Soc. Rock Mech., IIA, 580-582, 1974.

Worotnicki, G. and D. Denham, The state of stress in the upper parts of the Earth's crust in Australia according to measurements in mines and tunnels and from seismic observations, paper presented at the Symposium on Investigation of Stress in Rock, Int. Soc. for Rock Mech., Sydney, Australia, 1976.

Zoback, M.D., F. Rummel, R. Jung and C.B. Raleigh, Laboratory hydraulic fracturing experiments in intact and pre-fractured rock, Int. J. Rock Mech. Min. Sci. & Geomech. Abstr., 14, 49-58, 1977.

IMPROVED METHODS FOR MEASURING PRESENT CRUSTAL MOVEMENTS

Peter L. Bender[*]

Joint Institute for Laboratory Astrophysics, National Bureau of Standards and
University of Colorado, Boulder, Colorado 80309

Abstract. Improvements in geodetic measurement
techniques are likely to play an important role
in a number of types of geodynamics studies during the next decade. Increased accuracy for
horizontal distance and gravity measurements at
many sites is expected using mutliple-wavelength
measurements and falling-retroreflector gravimeters. Improved tiltmeters and strainmeters are
being developed, with attention being given to
decreasing perturbations due to very local ground
noise. Geodetic receivers using signals from the
Global Positioning System satellites probably
will make possible rapid relative position measurements with 1 to 3 cm accuracy in and around
seismic zones. An international program of
worldwide position measurements with about 3 cm
accuracy is planned, using both laser range
measurements to the LAGEOS satellite and long
baseline radio interferometry.

Introduction

Most of the subject matter covered in this
volume is limited to plate interiors. However,
the new space techniques for making geodetic measurements are likely to contribute strongly to
studies of large scale tectonic movements between
plates as well as within plates. In addition,
the classical geodetic techniques which have provided much of our information on tectonic distortions at plate boundaries also are used to study
local deformation and uplift in plate interiors.
To avoid duplication, it was decided to discuss
all of the geodetic measurement methods that are
likely to contribute strongly to geodynamics
studies in the present article, rather than re-
peating the discussions of some techniques in the
reports of different ICG Working Groups. Recent
developments in measurement techniques and the
prospects for further improvements in the next
few years will be emphasized.

So far, the direct determinations of strain
accumulation rates at plate boundaries and of
deformation related to seismic events have come
mainly from resurvey measurements using triangulation, trilateration, and leveling. Additional
results have been obtained from tide gages,
gravimeters, creepmeters, and tilt and strain
observations. Some of the results are discussed
in the report of ICG Working Group No. 10. Information on vertical movements in plate interiors obtained by leveling is discussed by Brown
and Reilinger elsewhere in this volume.

We will discuss first the ground measurement
techniques, and then the new space techniques.
For the ground techniques, we will concentrate on
modulated laser distance measurements, gravity
measurements, and strain and tilt measurements
using relatively long baselines. The space techniques include satellite laser ranging, long
baseline radio interferometry using signals from
astronomical sources, and radio interferometry
using signals from the Global Positioning System
satellites.

Improved Ground Techniques

A number of valuable articles concerning both
conventional and new measurement methods have
become available recently in three publications.
One is the Proceedings of the 1977 International
Symposium on Recent Crustal Movements [Whitten et
al., 1979], which appeared as a special volume of
Tectonophysics. The second is Application of
Geodesy to Geodynamics [Mueller, 1978], which is
the Proceedings of the Ninth Geodesy/Solid Earth
and Ocean Physics Research Conference. The third
is Proc. of Conf. VII, Measurement of Stress and
Strain Pertinent to Earthquake Prediction [Clark
et al., 1978]. In view of the availability of
these recent sources of information, I have
limited this article to discussions of the most
accurate methods for determining changes in
distance, tilt, strain, and gravity.

Electromagnetic Distance Measurements

The most accurate method at present for determining distances between points a number of kilometers apart is the use of laser beams modulated
at radio or microwave frequencies. In commer-

---

*Staff Member, Quantum Physics Division, NBS.

cially available instruments, the light goes out through a modulator, is reflected from a retro-reflector at the far end of the path, and then returns through the modulator again to a detector. The average signal from the detector is a maximum if the round-trip optical path length is an integral multiple of the wavelength of the modulating frequency, and thus the optical path length can be measured in terms of this wavelength. For the best instruments, the accuracy usually will be limited mainly by how well the atmospheric density along the path is known, and therefore how well the correction for the slowing down of light in air can be made. An uncertainty of 1 K in the average temperature along the path corresponds to an uncertainty of one part per million in the corrected distance, even if the atmospheric pressure is known perfectly. This is a typical accuracy if measurements of the atmospheric temperature, pressure, and humidity at the end points are used to determine the correction, but no additional precautions are taken.

A successful program of repeat survey measurements with higher accuracy has been achieved by Savage and co-workers in the U.S. by flying an aircraft or helicopter along the optical path to measure the atmospheric conditions. The repeatability of the measurements has been found to be given by the formula

$$\sigma = (a^2 + b^2 L^2)^{1/2}$$

where a = 3 mm, b = 2 × 10⁻⁷, and L is the path length [Savage and Prescott, 1973]. Roughly 800 lines/yr in the western U.S. have been measured in recent years. The percentage increase in cost for the airborne measurement is about 40%. Measurements with higher accuracy also have been achieved by Parm [1976] over quite level terrain in Finland by means of very careful experimental procedures, such as taking data only when the wind is blowing along the line of sight.

Another approach for improving the accuracy of the results has been developed by Slater and Huggett [1976] and has been demonstrated for a number of baselines near Hollister, California [Huggett et al., 1977; Slater, 1978; Slater and Burford, 1979]. Measurements are made with both red and blue lasers, and the difference in optical path length is used as a direct measure of the integrated atmospheric density along the path. Since the difference in distance for the wavelengths used is only about 5% of the total atmospheric effect on the distance, great care has to be taken to make the differential measurements as accurate as possible. However, the results indicate that the precision of the corrected distance is about one part in 10⁷ and the accuracy may be the same. Microwave distance measurements over nearly the same path sometimes are used to determine and correct for the integrated water vapor along the line of sight.

So far the multiple wavelength method has been considerably more limited in range than single-wavelength optical distance measurements. However, by using a three-wavelength system where the red and blue light beams and a microwave signal travel only one way over the path, it appears feasible to achieve an accuracy of 5 parts in 10⁸ for determining the atmospheric correction over paths as long as 40 or 50 km [Moody and Levine, 1979; Levine, 1978]. Such an atmospheric correction measuring device can be used with a separate instrument which determines the path length for red light, or this additional capability can be built into a single instrument. However, no test data with such instruments are available yet.

Gravity Measurements

One method of searching for evidence of vertical crustal movements is to make periodic remeasurements of the acceleration of gravity. While there are some situations where elevation changes can occur without variations in gravity, this probably is rare. The change in some important cases is roughly two to three microgals per centimeter of uplift. If both gravity changes and elevation changes can be measured, the combination provides an additional constraint on the physical mechanism responsible for the variations. Sites for gravity measurements have to be chosen carefully, since effects such as the withdrawal of water from aquifers can change the results. However, the choice of adequate sites on crystaline rock outcrops appears to be feasible in many non-sedimentary areas.

One type of instrument frequently used for gravity measurements is the LaCoste-Romberg gravimeter. By using several instruments and going back and forth between two sites, changes in the gravity difference can be detected at the level of roughly 10 or 15 microgals [Jachens, 1978a,b; Fett, 1978; Harrison and LaCoste, 1978; Brein et al., 1977; Kiviniemi, 1974]. Further instrumental improvements using electrical feedback may be feasible (W. E. Farrell, private communication, 1979). For comparing sites only a kilometer or two apart with nearly the same value of gravity, higher accuracy can be achieved [Lambert and Beaumont, 1977; Lambert, 1978; Lambert et al., 1979].

A number of networks for making repeated gravity measurements have been set up in different parts of the world. One of the interesting networks is one in southern California which includes about 300 sites with typical spacings of about 25 km, and 11 secondary reference sites which are carefully inter-compared [Jachens, 1978a; Jachens and Roberts, 1979].

During the past decade, substantial progress has been made on developing absolute gravimeters, which also may be quite valuable for crustal movement measurements. An instrument located in Paris has been operating for some time with an accuracy of a few microgals [Sakuma, 1974a,b] and another instrument with similar accuracy has been built in Japan. Transportable absolute gravimeters were first used to provide some of the reference points for the 1971 International Gravity Network. More recent transportable instruments have been developed in western Europe by joint Italian and French efforts [Cannizzo et al. 1978], in the U.S. [Hammond and Iliff, 1978], and in the U.S.S.R. [Boulanger, 1979]. The accuracy achieved is believed to be about 10 microgals. Work also is proceeding on a more portable absolute gravimeter which has an accuracy goal of 3 microgals [Faller et al., 1979].

A third type of gravimeter which has demonstrated very high stability in use at fixed sites is the superconducting gravimeter [Goodkind, 1979]. These devices have achieved a stability of a few microgals over several months [Goodkind, 1978]. This approaches the level at which effects such as vertical station displacements due to quasi-seasonal transport of water from the oceans to continental areas may become observable in some regions. The installation of superconducting gravimeters at tectonically interesting sites is one method of looking for both short term and intermediate term gravity variations. Such instruments also provide increased accuracy in the measurement of tidal gravity variations [Warburton and Goodkind, 1978; Goodkind, 1978].

Tilt and Strain

A substantial amount of work has been done in recent years on measurements of tilt with shallow borehole tiltmeters. However, it has become clear that a major limitation in such measurements is near-surface ground motions of meteorological or other non-tectonic origins [see e.g. various articles in Clark et al., 1978]. In order to improve the short-term stability, it seems necessary to have the points of support for the tiltmeters located further below the surface or much further apart. For improved long-term stability, either deeper borehole instruments or liquid level instruments with long baselines will be needed.

There are three main types of long baseline liquid tiltmeters in operation or under development at present. One uses a horizontal tube which is half-filled with liquid, and is free from systematic errors due to temperature variations along the tube [Beavan and Bilham, 1977; Plumb et al., 1978; Bilham et al., 1979]. However, the installation costs for very long instruments in irregular terrain may be high. The second type uses two tubes containing liquids with different thermal expansion coefficients [Huggett et al., 1976]. The effect of temperature variations then can be corrected for. A third type uses a measurement of differential pressure at the center of a thin tube connecting two end reservoirs of large diameter [Horsfall and King, 1978].

It is hoped that tests of the three types of long baseline liquid tiltmeters between the same end piers can be made soon in order to determine the instrumental accuracy. But research on pier stability also will be needed in order to find out what instrument length is required to obtain a given long-term stability. A substantial amount of data on fairly deep borehole tiltmeters is available [Cabaniss, 1978; Herbst, 1976; Akashi and Fukuo, 1977; Sato, 1979], but tests at the same site as the long baseline liquid instruments are highly desirable.

For strain measurements, many types of instruments have been developed. The most stable kind appear to be laser strainmeters [Berger and Wyatt, 1978; Berger and Levine, 1974; Levine, 1977; Beavan and Goulty, 1977; Seino et al., 1977]. However, even with a length of nearly a kilometer, the main stability limitation for instruments installed at the surface is the pier stability. Attempts are being made to overcome this problem by using optical anchors to measure pier motions with respect to reference points below the surface layers [Wyatt et al., 1979]. Research also is in progress with a new type of strainmeter using a bundle of carbon filaments as the reference [Hauksson et al., 1979] instead of a laser beam, an invar wire, or a fused silica tube.

Space Techniques

The basic approach used in the space techniques is to measure the distance or difference in distance from points on the ground to extraterrestrial reference points such as satellites or astronomical radio sources. The accuracy which appears to be achievable in such distance measurements is roughly 0.3 to 3 cm, and depends on many factors. For laser distance measurements, the inaccuracy is mainly due to uncertainty in the integrated atmospheric density along the line of sight and to the inadequacy of present procedures for measuring the round trip travel time. For radio measurements, the main problem is the uncertainty in the correction for the integrated amount of water vapor along the path. The effect of the ionosphere also has to be considered, but this can be corrected for accurately by comparing results of measurements made at two substantially different frequencies.

To be as specific as possible, the discussion will be limited to the following space techniques: laser range measurements to the Laser Geodynamics Satellite (LAGEOS); long baseline radio interferometry (LBI) using random noise emitted by extragalactic radio sources; and radio interferometry using signals from the Global Positioning System (GPS) satellites. The first two are expected to be used throughout the 1980's to measure tectonic plate motions and large scale distortions within the plates directly. The GPS approach is expected to be most useful for measurements at considerably larger numbers of points in and around seismic zones and at much shorter time intervals. Networks of continuously operating GPS receivers in seismic zones appear to be a real possibility, although the trade-offs with ground techniques in both accuracy and cost will have to be considered carefully. Mixed networks containing both improved ground instruments and GPS receivers may turn out to be desirable.

LAGEOS Ranging

The LAGEOS satellite [Smith et al., 1979a,b, Smith and Dunn, 1980] is in a nearly circular orbit with 110° inclination and 5900 km altitude. It is spherically symmetric and has a high density to minimize perturbations due to radiation pressure and atmospheric drag. About 15 ground stations currently are making range measurements to LAGEOS, which probably is at least as many as are needed to maintain good coverage of the orbit. However, changes in the locations of some of the ground stations would be quite helpful in improving the southern hemisphere coverage and in making use of sites with better weather, where the gaps in the data obtained would be considerably shorter.

The basic approach is to make range measurements to LAGEOS whenever possible from roughly 10 to 15 fixed or semi-fixed stations, and to determine the orbit and variations in the Earth's rotation from the data [Smith et al., 1979a]. Arc lengths of a week or so may be used in such fits. At suitable intervals, solutions involving a considerably larger amount of data can be carried out in order to obtain corrections to the station coordinates [Smith et al., 1979b], to some of the gravitational harmonic coefficients, to the ocean tide models [Eaves et al., 1979; Smith and Dunn, 1980], and to a few other parameters. Simulations [Bender and Goad, 1979] have indicated that such solutions can give accuracies of 4 or 5 cm for intercontinental distances, even without additional new information about the Earth's gravity field from other sources beyond that contained in NASA's GEM-10 gravity field model. Information on changes in the station positions should be considerably more accurate.

To obtain information on crustal movements at many more sites, a high-mobility LAGEOS ranging station is being constructed by the University of Texas [Silverberg, 1978], and two or three more probably will be built in Europe and the U.S. We can think of the orbit as being continually determined with respect to the fixed stations by range measurements from those sites. The high-mobility station then only has to obtain range measurements when LAGEOS is in three roughly perpendicular directions as seen from the ground in order to determine all three of its coordinates. In principle this could take only two satellite passes, but in practice four or more passes probably are needed. This is expected to take an average of one week or less per location for sites with 50% clear weather.

The main improvement required in LAGEOS ranging is in the basic measurement accuracy. At present many of the stations have 10 cm accuracy for the average residual over 100 returns, and the precision is considerably better. A few stations have 2 to 4 cm accuracy. What is needed is to upgrade roughly 10 of the stations to 1 cm basic measurement accuracy for 2 or 3 minute intervals as rapidly as possible.

Long Baseline Radio Interferometry

Although the radio noise arriving at the Earth from an extragalactic source is random in its amplitude and phase variations, it still can be used in high-accuracy distance measurements [see e.g. Counselman, 1976, 1980]. If two ground stations at the same site record the received signals for a fixed time, the cross-correlation between the two records will be maximum only if there is no time offset between them. If the stations are at different sites, the difference in travel time from the source to the two sites can be found by finding the delay time for one record with respect to the other that maximizes the cross correlation. The projection of the baseline onto the direction toward the source can be determined from the delay time, minus the clock difference between the two stations. For measurements with four or more sources in different directions in the sky, all three components of the baseline between the sites can be determined, as well as the clock difference.

A number of fixed stations already have been involved in accurate long baseline radio interferometry (LBI) measurements for geodynamics studies. These include five in the U.S., the Onsala Observatory in Sweden, the Effelsberg Observatory in the Federal Republic of Germany, and the NASA Deep Space Network stations in Spain and Australia. With the introduction of the new Mark III ground systems at a number of the sites, the accuracy of the results is expected to be limited mainly by the uncertainty in the tropospheric propagation velocity due to water vapor.

A network of three stations in the U.S. to monitor the Earth's rotation at least several times per week is being set up by the National Geodetic Survey [Carter et al., 1979; Carter, 1980]. Hopefully, this will become part of a worldwide system in the future. A number of countries have expressed interest in taking part in such a program. Encouraging results already are being obtained by several groups [Fanselow et al., 1979; Robertson et al., 1980]. The total number of stations required to monitor the Earth's rotation is smaller than for LAGEOS because measurements can be made even during cloudy weather or moderate rain. Thus about eight stations may be sufficient for this purpose, with some additional ones in the southern hemisphere needed part of the time to serve as reference points for crustal movement measurements.

As with LAGEOS ranging, it is expected that the large majority of crustal movement measurements will be made by high-mobility stations. The NASA Jet Propulsion Laboratory has operated a mobile

LBI station with a 9 m antenna at various sites in California during the past few years [Niell et al., 1979]. A new high-mobility station using a 4 m antenna will soon be in operation. The measurement time at each site is two days or less with either station, although the 9 m antenna takes substantially longer to move from site to site. If both stations were equipped with 4 m antennas, the number of sites which could be redetermined per year is quite large. Or, alternatively, the 9 m station can serve as a reference station for the 4 m station when it is operating at very remote sites where fixed reference stations cannot observe the same sources simultaneously.

The main area where improvements will be needed concerns the determination of the water vapor correction. The most promising approach is to use water vapor radiometers to measure the emission from water molecules along the line of sight. The observed power in a H_2O molecular emission line can be combined with an estimate of the average atmospheric temperature to give the integrated water vapor content of the atmosphere. Results obtained with several somewhat different types of radiometers have been reported by Guiraud et al. [1979], Moran and Rosen [1980], and Resch and Claflin [1980]. The most extensive results so far are those of Guiraud et al. [1979], but observations were reported only for vertical paths. Seven additional radiometers have been designed and assembled recently by the Jet Propulsion Laboratory for use in LBI measurements [Resch and Claflin, 1980].

There is a good theoretical basis for expecting that 1 cm accuracy can be achieved with water vapor radiometers, even at elevation angles as low as 20° [Westwater, 1978; Wu, 1979]. But direct measurements of radiometer performance for low elevation angles and under varying atmospheric conditions still are needed [Resch and Claflin, 1980], in support of both LBI and Global Positioning System measurement programs.

Measurements of baseline accuracy and reproducibility now are available for a number of LBI baselines. Results obtained with a 1.24 km baseline over a period of 15 months show rms variations of 3 mm, 5 mm, and 7 mm respectively in the baseline length, azimuth, and elevation [Rogers et al., 1978]. The mean results agree to 6 mm or better in each coordinate with the values obtained by careful ground surveying [Carter et al., 1980]. Measurements at different times with the 9 m mobile station at two sites separated by 42 km gave a baseline length which agreed with ground survey measurements to 6 ± 10 cm [Niell et al., 1979]. And measurements over a 4000 km baseline between the Haystack Observatory in Massachusetts and the Owens Valley Radio Observatory in California have given results with a 4 cm rms reproducibility over a two-year period [Robertson et al., 1979]. Recently, the first mesurements of intercontinental distances with

sub-decimeter reproducibility have been reported [Herring et al., 1980]. It is encouraging that the above results were obtained even without the use of water vapor radiometers.

Global Positioning System

The third space technique to be discussed involves the use of microwave signals at 1575 and 1227 MHz transmitted by the satellites of the Global Positioning System [Parkinson, 1979]. This System is being established by the U. S. Government for navigation purposes, and is intended to include either 18 or 24 satellites when it is completed in about 1987. The satellites are being placed in 12 hour orbits, with about 60° inclination and 20,000 km altitude. The first six satellites were in orbit by mid-1980.

Each of the signals transmitted by the GPS satellites is 180° phase modulated with a 10.23 MHz pseudo-random code, so that the power received is spread out over about a 20 MHz bandwidth and the carrier is suppressed [Spilker, 1978]. However, by using a similar code generator in the receiver on the ground, it is possible to produce a sinusoidal output frequency which will change its phase by one cycle each time the distance from the satellite to the receiver changes by one wavelength. This signal is called the reconstructed carrier, even though it is at a much lower frequency than the satellite transmission. By making nearly simultaneous phase measurements on the reconstructed carrier with two receivers, information on the projection of the baseline in the satellite direction can be obtained. By switching both of the receivers to a new satellite about every six seconds, and observing at least four satellites, the three components of the baseline and the difference of the receiver clocks can be determined [Bossler et al., 1980; Counselman et al., 1980].

An alternate approach is to treat the received signals from the GPS satellites as noise, and use the conventional LBI method [MacDoran, 1978, 1979]. The received power per unit bandwidth is about 10^5 times stronger than for extragalactic radio sources, so that much simpler ground equipment can be used. This approach has the advantage that knowledge of the code is not required. However, directional antennas with roughly 20 dB gain are needed in order to pick out which satellite is being observed.

The descriptions given above of the two main methods for using the GPS signals to determine crustal movements are very brief. However, the basic interferometric approach underlying both methods has been checked out thoroughly in astronomical LBI measurements [Counselman, 1976]. Both methods are expected to give accuracies limited mainly by the uncertainty in the water vapor corrections for short baselines. There also will be some error introduced by the GPS satellite orbit uncertainties. But this will be

only 1 cm per 100 km of baseline length for 2 m
orbit uncertainties, and accurate differential
tracking of the GPS satellites from 4 or more
well-separated ground stations at known locations
can reduce the orbit errors considerably further.

The main point to be made is that the use of
GPS signals with one of the two methods discussed
above is likely to give just as high accuracy for
short baselines as VLBI measurements with astronomical sources, and to be much simpler. GPS
methods thus are likely to be useful whenever
frequently repeated measurements are needed in
seismic zones. For example, nearly continuous
measurements over many baselines up to 100 km or
more in length will be possible after all of the
satellites are up, if they are needed.

It should be mentioned that the Doppler method,
which has been used extensively for geodetic measurements with signals from the U.S. Navy Naviga-
tional Satellite System, also can be used with
the GPS reconstructed carrier signals [Anderle,
1979]. By making differential measurements between two ground stations, the effect of satellite clock instability is removed. In addition,
the limitation because of instability in the
ground clocks can be removed by rapid switching
between the different satellites. However, the
apparatus required is essentially the same as for
the reconstructed carrier phase method, and the
accuracy is not likely to be as good.

One other method for using signals from the
GPS satellites has been suggested [Counselman and
Shapiro, 1979]. The approach would be to add low
power transmitters to each satellite which would
emit up to 10 sinusoidal signals at frequencies
between 1000 and 2000 MHz. With a number of
sinusoidal signals available, the receivers could
be much simpler than for the other methods discussed. However, it is not yet known whether it
will be possible to have the necessary transmitters added to future GPS satellites.

References

Akashi, K., and N. Fukuo, Seismometers and
strainmeters of borehole type, Seismitsu Kikai,
43, 111-118 (in Japanese), 1977.

Anderle, R.J., Accuracy of geodetic solutions
based on Doppler measurements of the NAVSTAR
Global positioning system satellites, Bull.
Geod. 53, 109-116, 1979.

Beavan, J., and R. Bilham, Thermally induced
errors in fluid tube tiltmeters, J. Geophys.
Res. 82, 5699-5704, 1977.

Beavan, R.J., and N.R. Goulty, Earth-strain
observations made with the Cambridge laser
strainmeter, Geophys. J. R. Astron. Soc. 48,
293-305, 1977.

Bender, P.L., and C.C. Goad, Probable LAGEOS
contributions to a worldwide geodynamics control network, in The Use of Artificial Satellites for Geodesy and Geodynamics, Vol. II, G.
Veis and E. Livieratos, eds., National Technical University, Athens, 145-161, 1979.

Berger, J., and J. Levine, The spectrum of earth
strain from 10^{-8} to 10^2 Hz, J. Geophys. Res.
79, 1210-1214, 1974.

Berger, J., and F. Wyatt, Some remarks on the
base length of tilt and strain measurements, in
Proc. of Conf. VII, Measurement of Stress and
Strain Pertinent to Earthquake Prediction,
USGS Open-File Report 79-370, Menlo Park,
California, 3-32, 1978.

Bilham, R., R. Plumb, and J. Beavan, Design
considerations in an ultra-stable, long baseline
tiltmeter -- results from a laser tiltmeter, in
Terrestrial and Space Techniques in Earthquake
Prediction Research, A. Vogel, ed., Friedr.
Vieweg & Sohn, Braunschweig/Wiesbaden, 235-254,
1979.

Bossler, J.D., C.C. Goad, and P.L. Bender, Using
GPS for geodetic positioning, Bull. Geod.,
submitted, 1980

Boulanger, J.D., Certain results of absolute
gravity determinations by the instrument of the
U.S.S.R. Academy of Sciences, in Publication
Dedicated to T. J. Kukkamäki on the Occasion of
his 70th Anniversary, J. Kakkuri, ed., Pub.
No. 89 of the Finnish Geodetic Institute,
Helsinki, 20-26, 1979.

Brein, R., C. Gerstenecker, A. Kiviniemi, and L.
Petterson, Report on high precision gravimetry,
Prof. Pap. Natl. Land Surv., Gavle, Sweden,
1977.

Cabaniss, G.H., The measurement of long period
and secular deformation with deep borehole
tiltmeters, in Applications of Geodesy to
Geodynamics, I.I. Mueller, ed., Report No.
280, Dept. of Geodetic Science, Ohio State
University, 165-169, 1978.

Cannizzo, L., G. Cerutti, and I. Marson,
Absolute-gravity measurements in Europe, Il
Nuovo Cimento, 1C, 39-85, 1978.

Carter, W.E., Project POLARIS: a status report,
in Radio Interferometry Techniques for Geodesy,
NASA Conf. Pub. 2115, Wash., D.C., 455-460,
1980.

Carter, W.E., D.S. Robertson, and M.D. Abell, An
improved polar motion and Earth rotation monitoring service using radio interferometry, in
IAU Symp. No. 82, Time and the Earth's Rotation, D.D. McCarthy and J.D.H. Pilkington, eds.,
D. Reidel, Dordrecht, Holland, 191-197, 1979.

Carter, W.E., A.E.E. Rogers, C.C. Counselman III,
and I.I. Shapiro, Comparison of geodetic and
radio interferometric measurements of the
Haystack-Westford baseline vector, J. Geophys.
Res., 85, 2685-2687, 1980.

Clark, B.R., J.H. Pfluke, and J.F. Evernden,
eds., Proc. of Conf. VII, Measurement of Stress
and Strain Pertinent to Earthquake Prediction,
USGS Open-File Report 79-370, Menlo Park,
California, 1978.

Counselman, C.C. III, Radio astronomy, Ann. Rev.
Astron. Astrophys. 14, 197-214, 1976.

Counselman, C.C. III, Meeting on radio interferometric techniques for geodesy, Trans. AGU
60, 673-674, 1979. See also papers in Radio

Interferometry Techniques for Geodesy, loc.
cit., 1980.

Counselman, C.C. III, and I.I. Shapiro, Minia-
ture interferometric terminals for earth surveying, Bull. Geod. 53, 139-163, 1979.

Counselman, C.C. III, I.I. Shapiro, R.L.
Greenspan, and D.B. Cox, Jr., Backpack VLBI
terminal with subcentimeter capability, in
Radio Interferometry Techniques for Geodesy,
loc. cit., 409-414, 1980.

Eaves, R.J., B.E. Schutz, and B.D. Tapley, Ocean
tide perturbations on the LAGEOS orbit (abst.),
Trans. AGU 60, 808, 1979.

Faller, J.E., R.L. Rinker, and M.A. Zumberge,
Plans for the development of a portable absolute gravimeter with a few parts in 10^9 accuracy, Tectonophys. 52, 107-116, 1979.

Fanselow, J.L., J.B. Thomas, E.J. Cohen, P.F.
MacDoran, W.G. Melbourne, B.D. Mulhall, G.H.
Purcell, D.H. Rogstad, L.J. Skjerve, D.J.
Spitzmesser, J. Urech, and G. Nicholson,
Determination of UT1 and polar motion by the
Deep Space Network using very long baseline
interferometry, in IAU Symp. No. 82, Time and
the Earth's Rotation, loc. cit., 199-209,
1979.

Fett, J.D., Periodic high precision gravity observations in Southern California, in Proc. of
Conf. VII, Measurement of Stress and Strain
Pertinent to Earthquake Prediction, loc. cit.,
158-161, 1978.

Goodkind, J.M., High precision tide spectroscopy,
in Applications of Geodesy to Geodynamics, loc.
cit., 309-311, 1978.

Goodkind, J.M., Continuous measurements with the
superconductivity gravimeter, Tectonophys. 52,
99-105, 1979.

Guiraud, F.O., J. Howard, and D.C. Hogg, A dual-
channel microwave radiometer for measurement of
precipitable water vapor and liquid, IEEE Trans.
on Geoscience Elect. GE-17, 129-136, 1979.

Hammond, J.A., and R.L. Iliff, The AFGL absolute
gravity program, in Applications of Geodesy to
Geodynamics, loc. cit., 245-254, 1978.

Harrison, J.C., and L.J.B. LaCoste, The measurement of surface gravity, in Applications of
Geodesy to Geodynamics, loc. cit., 239-244,
1978.

Hauksson, E., J. Beavan, and R. Bilham, Improved
carbon-fiber extensometers (abst.), Trans. AGU
60, 936, 1979.

Herbst, K., Interpretation of Tilt Measurements
in the Period Range Above that of the Tides,
Ph.D. thesis, Tech. University of Clausthal,
Federal Republic of Germany, 1976.

Herring, T.A., B.E. Corey, C.C. Counselman III,
I.I. Shapiro, B.O. Rönnäng, O.E.H. Rydbeck,
T.A. Clark, R.J. Coates, C. Ma, J.W. Ryan, N.R.
Vandenberg, H.F. Hinteregger, C.A. Knight,
A.E.E. Rogers, A.R. Whitney, D.S. Robertson,
and B.R. Schupler, Geodesy by radio inter-
ferometry: Intercontinental distance deter-
minations with subdecimeter precision, J.
Geophys. Res., submitted, 1980.

Horsfall, J.A.C., and G.C.P. King, A new geophysical tiltmeter, Nature 274, 675-676, 1978.

Huggett, G.R., L.E. Slater, and G. Pavlis, Precision leveling with a two-fluid tiltmeter,
Geophys. Res. Lett. 3, 754-756, 1976.

Huggett, G.R., L.E. Slater, and J. Langbein,
Fault slip episodes near Hollister, California:
Initial results using a multiwavelength
distance-measuring instrument, J. Geophys. Res.
82, 3361-3368, 1977.

Jachens, R.C., Temporal gravity changes as applied to studies of crustal deformation, in
Proc. of Conf. VII, Measurement of Stress and
Strain Pertinent to Earthquake Prediction, loc.
cit., 222-243, 1978a.

Jachens, R.C., The gravity method and interpre-
tive techniques for detecting vertical crustal
movements, in Applications of Geodesy to
Geodynamics, loc. cit., 153-155, 1978b.

Jachens, R.C., and C.W. Roberts, Gravity stable
in southern California, 1976-1978 (abst.),
Trans. AGU 60, 810-811, 1979.

Kiviniemi, A., High precision measurements for
studying the secular variations in gravity in
Finland, Pub. No. 78 of the Finnish Geodetic
Institute, 1974.

Lambert, A., Review of Canadian experience in
precise gravimetry, in Applications of Geodesy
to Geodynamics, loc. cit., 157-158, 1978.

Lambert, A., and C. Beaumont, Nano variations in
gravity due to seasonal groundwater movements:
Implications for the gravitational detection of
tectonic movements, J. Geophys. Res. 82, 297-
306, 1977.

Lambert, A., J. Liard, and H. Dragert, Canadian
precise gravity networks for crustal movement
studies: An instrument evaluation, Tectonophys.
52, 87-96, 1979.

Levine, J., Laser distance-measuring techniques,
Ann. Rev. Earth Planet. Sci. 5, 357-369, 1977.

Levine, J., Multiple wavelength geodesy, in
Applications of Geodesy to Geodynamics, loc.
cit., 99-102, 1978.

MacDoran, P.F., SERIES GPS geodetic system
(abst.), Trans. AGU 59, 1052, 1978.

MacDoran, P.F., Satellite emission radio interferometric earth surveying SERIES-GPS geodetic
system, Bull. Geod. 53, 117-138, 1979.

Moody, S.E., and J. Levine, Design of an extended
range three wavelength distance measuring instrument, Tectonophys. 52, 77-82, 1979.

Moran, J.M., and B.R. Rosen, The estimation of
the propagation delay through the troposphere
from microwave radiometer data, Radio Science,
submitted, 1980.

Mueller, I.I., ed., Applications of Geodesy to
Geodynamics, loc. cit., 1978.

Niell, A.E., K.M. Ong, P.F. MacDoran, G.M. Resch,
D.D. Morabito, E.S. Claflin, and J.F. Dracup,
Comparison of a radio interferometric differential baseline measurement with conventional
geodesy, Tectonophys. 52, 49-58, 1979.

Parkinson, B.W., The Global Positioning System
(NAVSTAR), Bull. Geod. 53, 89-108, 1979.

Parm, T., High Precision Traverse of Finland, Finnish Geodetic Institute, Publ. No. 79, 108 pp., Helsinki, 1976.

Plumb, R., R. Bilham, and J. Beavan, A stable long baseline fluid tiltmeter for tectonic studies, in Proc. of Conf. VII, Measurement of Stress and Strain Pertinent to Earthquake Prediction, loc. cit., 47-83, 1978.

Resch, G.M., and E.S. Claflin, Microwave radiometry as a tool to calibrate tropospheric water-vapor delay, in Radio Interferometry Techniques for Geodesy, loc. cit., 377-384, 1980.

Robertson, D.S., W.E. Carter, B.E. Corey, C.C. Counselman III,, I.I. Shapiro, J.J. Wittels, H.F. Hinteregger, C.A. Knight, A.E.E. Rogers, A.R. Whitney, J.W. Ryan, T.A. Clark, R.J. Coates, C. Ma, J.M. Moran, Recent results of radio interfereometric determinations of a transcontinental baseline, polar motion, and earth rotation, in IAU Symp. No. 82, Time and the Earth's Rotation, loc. cit., 217-224, 1979.

Robertson, D.S., T.A. Clark, R.J. Coates, C. Ma, J.W. Ryan, B.E. Corey, C.C. Counselman, R.W. King, I.I. Shapiro, H.F. Hinteregger, C.A. Knight, A.E.E. Rogers, A.R. Whitney, J.C. Pigg, and B.R. Schupler, Polar motion and UT1: Comparison of VLBI, lunar laser, satellite laser, satellite Doppler, and conventional astrometric determinations, in Radio Interferometry Techniques for Geodesy, loc. cit., 33-44, 1980.

Rogers, A.E.E., C.A. Knight, H.F. Hinteregger, A.R. Whitney, C.C. Counselman III, I.I. Shapiro, S.A. Gourevitch, and T.A. Clark, Geodesy by radio interferometry: Determination of a 1.24-km base line vector with ~5 mm repeatability, J. Geophys. Res. 83, 325-334, 1978.

Sakuma, A., Report on absolute measurement of gravity, in Proceedings of the International Symposium on the Earth's Gravitational Field and Secular Variations in Position, Dept. of Geodesy, Univ. of New South Wales, Sydney, 674-684, 1974a.

Sakuma, A., Report on absolute measurements of gravity, in Bull. d'Information No. 35, Bureau Gravimetrique International, I-39-I-42, 1974b.

Sato, H., A short note on borehole-type tiltmeters and earthquake prediction, Res. Note National Research Center for Disaster Prevention, 34, 32 pp., 1979.

Savage, J.C., and W.H. Prescott, Precision geodetic distance measurements for determining fault movements, J. Geophys. Res. 78, 6001-6008, 1973.

Seino, S., T. Ohishi, and Y. Sakurai, Measurement of earth strains with a laser interferometer, Trans. Soc. Instrum. Control Engrs. 13, 174-179, 1977.

Silverberg, E.C., Mobile satellite ranging, in Applications of Geodesy to Geodynamics, loc. cit., 41-46, 1978.

Slater, L.E., Crustal deformation and aseismic fault slip near Hollister, California, in Proc. of Conf. VII, Measurement of Stress and Strain Pertinent to Earthquake Prediction, loc. cit., 502-520, 1978.

Slater, L.E., and R.O. Burford, A comparison of long-baseline strain data and surface fault creep near Hollister, California, Tectonophys. 52, 481-496, 1979.

Slater, L.E., and G.R. Huggett, A multiwavelength distance measuring instrument for geophysical experiments, J. Geophys. Res. 81, 6299-6305, 1976.

Smith, D.E., and P.J. Dunn, Long term evolution of the LAGEOS orbit, Geophys. Res. Lett., 7, 437-440, 1980.

Smith, D.E., R. Kolenkiewicz, P.J. Dunn, and M. Torrence, Determination of polar motion and Earth rotation from laser tracking of satellites, in IAU Symp. No. 82, Time and the Earth's Rotation, loc. cit., 247-255, 1979a.

Smith, D.E., R. Kolenkiewicz, P.J. Dunn, and M.H. Torrence, Determination of station coordinates from LAGEOS, in The Use of Artificial Satellites for Geodesy and Geodynamics, Vol. II, loc. cit., 162-172, 1979b.

Spilker, J.J., Jr., GPS signal structure and performance characteristics, Navigation 25, 121-146, 1978.

Warburton, R.J., and J.M. Goodkind, Detailed gravity tide spectrum between 1 and 4 cycles per day, Geophys. J. R. Astron. Soc. 52, 117-136, 1978.

Westwater, E.R., The accuracy of water vapor and cloud liquid determination by dual-frequency ground-based microwave radiometry, Radio Science, 13, 677-685, 1978.

Whitten, C.A., R. Green, and B.K. Meade, eds., Tectonophys. 52 (Special Issue), 1979.

Wu, S.-C., Optimum frequencies of a passive microwave radiometer for tropospheric pathlength correction, IEEE Trans. Ant. Prop., AP-27, 233-239, 1979.

Wyatt, F., J. Berger, and K. Beckstrom, Precision measurements of surface benchmarks (abst.), Trans. AGU 60, 811, 1979.